FSMA and Food Safety Systems

FSMA and Food Safety Systems

Understanding and Implementing the Rules

Jeffrey T. Barach

Barach Enterprises, LLC

This edition first published 2017 © 2017 by John Wiley & Sons, Ltd.

Registered Office
John Wiley & Sons, Ltd, The Atrium, Southern Gate, Chichester, West Sussex, PO19 8SQ, UK

Editorial Offices
9600 Garsington Road, Oxford, OX4 2DQ, UK
The Atrium, Southern Gate, Chichester, West Sussex, PO19 8SQ, UK
111 River Street, Hoboken, NJ 07030-5774, USA

For details of our global editorial offices, for customer services and for information about how to apply for permission to reuse the copyright material in this book please see our website at www.wiley.com/wiley-blackwell.

Library of Congress Cataloging-in-Publication data applied for

ISBN: 9781119258070

A catalogue record for this book is available from the British Library.

Wiley also publishes its books in a variety of electronic formats. Some content that appears in print may not be available in electronic books.

Cover image: Background: AdShooter/Getter Images; Smarties: Foodio/Shutterstock;
Ice cream: ASA studio/Shutterstock; Cookies: Goran Bogicevic/Shutterstock;
Vegetables: margouillat photo/Shutterstock

Set in 10/12pt Warnock by SPi Global, Pondicherry, India

10 9 8 7 6 5 4 3 2 1

Contents

Preface

The concept of Food Safety Systems to control food hazards has been advancing over the past twenty-five years with the global adoption of Hazard Analysis Critical Control Point (HACCP) principles. This has been both on a voluntary industry basis and as a mandatory regulation in the United States for seafood, juice, and meat & poultry products. Prior approaches of in-plant inspections and end-product testing have proven of limited value and generally result in reacting to problems, rather than being proactive to prevent problems from occurring. Recently, we see that FDA has been building upon the basic principles and practices of HACCP with a further refined and more comprehensive Food Safety System outlined in the final rules for "Current Good Manufacturing Practice, Hazard Analysis, and Risk-Based Preventive Controls for Human Food (Federal Register (80) September 17, 2015). The final rule for human foods is the subject of this book; however, the Food Safety Modernization Act of 2011 (FSMA) is quite extensive. Since January 2013, FDA has proposed seven foundational rules to implement FSMA. Those rules became final in 2015 and 2016: Preventive Controls for Human Food, Preventive Controls for Animal Food, Produce Safety, Foreign Supplier Verification Program, Third Party Certification, Sanitary Transportation, and Intentional Adulteration. The goal of industry and regulators is to have a Food Safety System that is built upon a corporate food safety culture, a sound foundation of Good Manufacturing Practices (GMPs) prerequisite programs, and incorporates risk-based preventive controls directed at controlling hazards.

If your business already has HACCP in place, the transition to the human foods rules of the Food Safety Modernization Act of 2011will be relatively straightforward. If you currently have only GMPs in place, the upgrade will require more time and resources to make the change. This guide will help you in either case by guiding you through what is required and developing a Food Safety Plan. The guide also addresses managing the plan and confirming its effectiveness. Within the guide, you will find both information you need to explain how a Food Safety System works and what you must do to develop and properly administer it. You will also find forms you may use to help structure your own Food Safety Plan and many examples of things to consider in addressing food safety hazards. Several example Food Safety Plans are included, since a main purpose of this guide is designed to teach by example.

Developing and implementing your Food Safety System according to the rules of FSMA is not as difficult as you might imagine. Hopefully, as you read the material presented here and follow along with the examples, you will gain the knowledge you need.

Although the recommendations in this publication are based upon scientific literature, regulatory guidance, and wide industry experience, examples of Food Safety Plans and Food Safety System components are not to be construed as a guarantee that they are sufficient to prevent damage, spoilage, loss, accidents, or injuries resulting from the use of this information. Furthermore, the study and use of this publication by any person or company is not an assurance that a person or company is then proficient in the operations and procedures discussed in this publication. The use of the examples, statements, recommendations, or suggestions contained, herein are not to be considered as creating any responsibility for damage, spoilage, loss, accident, or injury resulting from such use.

About the Author

Jeffrey T. Barach, PhD; Principal, Barach Enterprises, LLC

Jeffrey Barach is a food scientist who has been active in research and development, regulatory liaison activities, teaching, problem solving, and trouble shooting for the food industry for over thirty years. He routinely participates as a consultant in planning, development, and management of special projects and programs for the food industry on health and safety issues, production of foods, regulatory compliance issues, and training. He is a subject matter expert (SME) on issues related to food safety modernization, biotechnology, food irradiation, nanotechnology, and other new processing and testing technologies. Dr. Barach was formerly with Grocery Manufacturers Association as head of their Science Policy Group and previously directed the GMA Laboratory in Washington, D.C. His education includes receiving his MS and PhD in food science from North Carolina State University and his undergraduate degree in Chemistry from the University of North Carolina. Dr. Barach is also a lecturer, lead instructor training professional and educator in food safety and food sanitation, with a focus on HACCP and FSMA food safety systems.

The author has met and worked with many outstanding food scientists from industry, academia, and the regulatory agencies in his career. He is grateful to those that shared their expertise, wisdom, and knowledge with him. He also is pleased to share his knowledge, as a teacher, with his many students and fellow colleagues. Hopefully, this guide book will extend that desired ambition of knowledge sharing. Special thanks goes to his wife Kathy and to those that helped him with the review of this publication. Their encouragement and suggestions are kindly taken and greatly appreciated.

1

What is Modern Food Safety, and How is that Different from HACCP?

1.1 Introduction

President Obama signed the Food Safety Modernization Act of 2011 (FSMA) in response to Congress' efforts to address the food safety issue demands of a broad coalition of stakeholders – including produce growers, food processors, retailers, and consumers who were disturbed by a series of illness outbreaks and contamination incidents. These significant foodborne illness events, which involved both domestic and imported foods, undermined consumer confidence and imposed harmful and costly disruptions on consumers and food producers. Many believed that these events would be largely preventable if new laws were developed and implemented that utilized best practices for preventing food safety problems.

The bills and subsequent law was focused on making these practices the norm for all domestic FDA-regulated products in the United States (U.S.) market. Congress and the FDA realized that a significant portion of the food consumed in the U.S. is produced in foreign countries. They recognized the need to address food safety at a global scale to address the wide range of food, ingredients, and commodities the U.S. imports from over 200 countries and territories.

For both domestic and foreign food, FSMA regulatory mandates keyed in on two proven basic principles of food safety that could address the concerning incidents of food safety problems and the growing diversity of the global food supply. The new rules were, in principle, to be a scientific assessment of risk and were to implement controls that would prevent significant food safety problems, rather than just to react to them after the fact.

FSMA regulations (Fed. Reg., 2015) explicitly recognize the food safety role of the food industry while giving regulatory authority to the FDA. With that, it rests on a third core principle that empowers those who produce

FSMA and Food Safety Systems: Understanding and Implementing the Rules,
First Edition. Jeffrey T. Barach.

food for the commercial market to have the responsibility and capability to make it safe in accordance with recognized best practices for preventing harmful contamination and preventing food safety hazards.

FSMA uses a risk-based strategy and preventive controls to achieve its broad goals by fundamentally changing FDA's food safety role and redefining its relationship with other participants in the food system. These broad steps toward modernizing food safety are fundamentally sound and inherently necessary in the global environment and under existing resource limitations if food safety goals are to be achieved.

The language of FSMA is consistent with the Hazard Analysis Critical Control Points (HACCP) approach. In the U.S., HACCP is required for many foods, including meat and poultry, seafood and juice (NACMCF, 1997; see Appendix B). Regulations within the European Economic Community require HACCP plans. Likewise, the Codex Alimentarius Commission notes that HACCP is a tool to assess hazards and establish control systems that focus on prevention rather than relying mainly on end-product testing. The keystone of FSMA, like HACCP, is the development of risk-based preventive controls for food facilities. Food facilities are required to develop and implement a written plan for preventive controls and a written recall plan.

This involves:

1) Evaluating the hazards that could affect food safety;
2) Specifying what preventive steps, or controls, will be put in place to significantly minimize or prevent the hazards;
3) Specifying how the facility will monitor these controls to ensure they are working;
4) Maintaining routine records of the monitoring, and
5) Specifying what actions the facility will take to correct problems that arise.

The final rule implements the requirements of FSMA for covered facilities to establish and implement a food safety system that includes sound sanitation programs, a hazard analysis, and risk-based preventive controls. Specifically, the rule establishes requirements for:

- A written food safety plan;
- Hazard analysis;
- Preventive controls;
- Monitoring;
- Corrective actions and corrections;
- Verification;
- Supply-chain program;
- Recall plan; and
- Associated records.

However, in order for the food safety plan to be effective, it must be built on a strong foundation of current good manufacturing practices (cGMPs). FSMA addresses cGMPs in the new rules under the Subpart B provisions. The importance of having effective and well implemented cGMPs can't be overstated. Many of the problems associated with low-moisture foods, such as peanut butter, that lead to foodborne illnesses were because of poor basic sanitation practices.

1.2 FSMA Sanitation and cGMPs

Several steps to modernize cGMPs have been incorporated in FSMA rules. For a better understanding of these developments, it may be helpful to recap where these initiatives originated. As background, the FDA issued a white paper in 2005, titled "Food cGMP Modernization – A Focus on Food Safety," to address significant changes that had occurred both in industry and in the science and technology of food safety. The last time cGMPs were revised was in the mid 1980s. As FSMA regulations were being formulated by FDA, it became apparent that new food safety regulations should combine this earlier cGMP effort with the new initiative that was focused on preventive controls. This strategy led to the rewrite of 21 CFR Part 110 into 21 CFR Part 117 (Subpart B). Rather than having FDA pursue further modernization of cGMPs as a separate task, almost all stakeholders agreed with incorporating those improvement activities within the FSMA framework. As a result, the new FSMA rules for human foods are titled "Current Good Manufacturing Practice, Hazard Analysis, and Risk-Based Preventive Controls for Human Food" to included both cGMP and preventive control updates.

Under the FSMA framework, cGMPs are revised and industry responsibilities are clarified. The rule states what must be done in a facility to control sanitation, and the language in the regulation was updated, such as using the term "must" instead of "shall." As with HACCP programs, requirements under cGMPs are included for personnel, plant and grounds, sanitary operations, sanitary facilities and controls, equipment and utensils, processes and controls, warehousing, and distribution. The cGMPs were modified to clarify that certain provisions requiring protection against contamination of food also require protection against allergen cross-contact. Regulations also now require cleaning of non-food-contact surfaces as frequently as necessary to protect against contamination of food and food-contact surfaces. Additionally, food-contact surfaces used for manufacturing/processing or holding low-moisture food must be in a clean, dry, and sanitary condition at the time of use. The rules place emphasis on education and training to ensure employees have the knowledge and/or experience necessary to make and produce safe food.

The following is a summary of the major components of FSMA regulations regarding cGMPs:

21 CFR Part 117 (Subpart B) – Current Good Manufacturing Practices
Part 117.10 Personnel Part 117.20 Plant and grounds Part 117.35 Sanitary operations Part 117.37 Sanitary facilities and controls Part 117.40 Equipment and utilities Part 117.80 Processes and controls Part 117.93 Warehousing and distribution Part 117.95 Holding and distribution of human food by-products for use as animal food Part 117.110 Defect action levels
Note: Additional regulations may be applicable, see final rules.

1.3 FSMA Preventive Controls

HACCP is the predecessor of FSMA. Both approaches have the goal of establishing a food safety system that will provide the environment for safe food production. That includes cGMPs and controls to prevent the food product from becoming adulterated by any known or foreseeable hazards coming from the foods or hazards resulting from faulty manufacturing operations. Contamination of food products typically comes from one of three different sources: 1) ingredients, 2) the processing environment, including equipment, or 3) people. A major difference between FSMA and HACCP is that HACCP is focused mainly on processing controls (called critical control points) while FSMA expands the aspects of controls to include allergen preventive controls, sanitation preventive controls, and supplier controls, as well as maintaining the importance process controls.

FSMA focuses on identifying hazards that are then mitigated by specific types of preventive controls. It requires a written analysis of potential hazards to determine if they are known or reasonably foreseeable and if they are severe enough to require a preventive control to be implemented. The first part of hazard analysis is the identification of potential biological, chemical (including radiological), and physical hazards that may be associated with the facility or the food. These hazards may occur naturally, may be unintentionally introduced, or may be intentionally introduced for economic gain. Controls are needed to manage the hazards introduced from these sources. With the significant hazards identified, preventive controls (of the types described above) are assigned and managed by plant personnel through well developed and documented actions and procedures. FSMA also requires a written Food Safety

Plan, and if hazards are identified that need preventive controls, they are included in the plan. The Food Safety Plan is the primary document that guides the preventive controls of the food safety system.

FSMA details the requirements and contents of the Food Safety Plan:

Part 117.126 Food Safety Plan requirements.
Prepare, or have prepared, and implement a Food Safety Plan. It must be prepared, or its preparation overseen, by one or more preventive controls qualified individuals. The written Food Safety Plan must include the written: • Hazard analysis • Preventive controls • Supply-chain program • Recall plan • Procedures for monitoring the implementation of the preventive controls • Corrective action procedures • Verification procedures, and • Records Note: Additional regulations may be applicable, see final rules.

1.4 Process Controls

Process controls for biological hazards depend on the organism(s) of public health significance, the characteristics of the food, and the process used to product the food. In many foods that have a high-moisture environment, pathogenic vegetative cells like *E. coli* O157:H7 and *Salmonella* are fairly easily killed during processing by heating or cooking. In low-moisture foods, however, *Salmonella* has unique thermal-resistance and durability characteristics that need to be carefully considered in the context of developing the food safety controls where *Salmonella* can survive. For pathogens that produce bacterial spores that can survive heat and outgrow, producing toxic food, a different control approach is necessary. If the spore-forming organism of public health concern is *Clostridium botulinium*, FSMA recognizes that well established regulations already exist in 21 CFR Part 113 and an exemption allows low-acid canners to control this biological hazard using those regulations (see Part 177.5). Other toxin-producing spore formers, such a *Clostridium perfringens* and *Bacillus cereus* require a different control approach, usually a cooling temperature control. When the hazard analysis indicates a pathogen hazard, there are many factors in deciding the appropriate process control. The type of organism (vegetative cell or spore), the nature and characteristics of the pathogen, the condition of the food product (e.g., moisture, pH, a_w) and other factors need to be considered. This may not be a simple task, given all the factors and variables involved. Often designing the proper process control may require the input from subject-matter experts such as a microbiologist, a process authority, a consultant and/or other food scientists.

In HACCP terminology, using a processing step to kill vegetative pathogens would typically be considered a critical control point (CCP). In FSMA, the concept is the same, and the action is called a process control. The potential hazard from bacterial pathogens and the appropriate mitigating steps would be identified in the hazard analysis with the oversight of the Preventive Control Qualified Individual (PCQI). Preventive controls would then become part of the Food Safety Plan as a process control, if a processing step is used to inactivate the pathogen. For example, a cooking process may be part of a high-moisture food manufacturing operation. With proper validation, it could be scientifically shown that this heating step will control *Salmonella* and remove the hazard. The cooking step becomes the point of control to be managed to ensure a *Salmonella* hazard does not exist for the food product.

When incorporating process controls into the Food Safety Plan, the preventive controls required include only those appropriate to the facility and the food, as determined by hazard analysis. Process controls would typically be at CCPs, similar to HACCP.

Part 117.135 (c)(1) Process controls.
Process controls include procedures, practices, and processes to ensure the control of parameters during operations such as heat processing, acidifying, irradiating, and refrigerating foods. Process controls must be included, as appropriate to the nature of the applicable control and its role in the facility's food safety system: i) Parameters associated with the control of the hazard; and ii) The maximum or minimum value, or combination of values, to which any biological, chemical, or physical parameter must be controlled to significantly minimize or prevent a hazard requiring a process control. Note: Additional regulations may be applicable, see final rules.

1.5 Sanitation Controls

Like process controls, sanitation preventive controls are those determined through a hazard analysis as necessary to significantly minimize or prevent: 1) environmental pathogens in a ready-to-eat (RTE) food exposed to the environment prior to packaging where the packaged food does not receive a treatment that would significantly minimize the pathogen; 2) biological hazards in a RTE food due to employee handling; and 3) food allergen hazards. Other aspects of routine (cGMP) sanitation such as pest control and safety of water and employee health do not need to be in a Food Safety Plan unless these programs/procedures are determined to be of a nature where hazards will result unless a preventive control is applied.

Part 117.135 (c)(3) Sanitation controls.

Sanitation controls include procedures, practices, and processes to ensure that the facility is maintained in a sanitary condition adequate to significantly minimize or prevent hazards such as environmental pathogens, biological hazards due to employee handling, and food allergen hazards. Sanitation controls must include, as appropriate to the facility and the food, procedures, practices, and processes for the:

i) Cleanliness of food-contact surfaces, including food-contact surfaces of utensils and equipment;
ii) Prevention of allergen cross-contact and cross-contamination from insanitary objects and from personnel to food, food packaging material, and other food-contact surfaces and from raw product to processed product.

Note: Additional regulations may be applicable, see final rules.

In all cases, effective sanitation procedures are a front-door to back-door necessity to ensure safe food production. A hazard may be re-introduced in an open plant environment postprocess. Routine plant sanitation standard operating procedures and specific sanitation preventive controls work together to ensure food safety.

FSMA identifies certain ready-to-eat (RTE) foods where environmental monitoring would be appropriate as a verification tool to confirm the effectiveness of the sanitation control. Foods such as peanut butter, dried dairy products for use in RTE foods, and roasted nuts are among the products for which manufacturing operations would need to have an environmental monitoring program when such foods are exposed to the environment. However, FSMA qualifies the need for implementing environmental monitoring as a possible verification activity as being appropriate to the food, facility, nature of the preventive control, and the role of that control in the facility's food safety system. Environmental monitoring generally would be required if contamination of a RTE food with an environmental pathogen is a hazard requiring a preventive control. If environmental monitoring is used as a verification tool for sanitation controls, the following procedures are recommended by FDA:

Environmental Monitoring Procedures

- Identify test microorganism
- Identify locations (may be guided by zoning*) and number of sites to be tested
- Identify timing and frequency for collecting and testing samples
- Identify test, including the analytical method
- Identify the laboratory conducting testing
- Include corrective action procedures

*(For information about zones and environmental monitoring, see 78 Federal Register 3646 at 3816).

Part 117.165 (a)(3) Verification – Environmental Monitoring
Environmental monitoring, for an environmental pathogen or for an appropriate indicator organism, if contamination of a RTE food with an environmental pathogen is a hazard requiring a preventive control, by collecting and testing environmental samples.
Note: Additional regulations may be applicable, see final rules.

1.6 Supplier Controls

Globalization of the food supply has brought new challenges to the food industry. Fortunately, many of the same practices and procedures used for managing foods and ingredients from domestic suppliers also apply to foreign suppliers. When an identified bacterial hazard is mitigated and controlled within the facility, the burden of controlling the pathogen is taken in-house. However, when a facility relies on their supplier to control the hazard: for example, if the food operation does not include a thermal processing step, the Food Safety Plan may need to include a supplier control to prevent pathogen-contaminated source material from entering the plant's operations. Also, when a potential pathogen is passed on to a customer, that customer needs to be alerted to that fact and take proper actions to control the hazard.

Supply-chain controls, implemented through a supply-chain program, are required for ingredients or raw materials for which the receiving facility's hazard analysis identified a hazard requiring a supply-chain-applied control. Other preventive controls may be identified as appropriate based on the hazard analysis. FSMA has introduce a regulation for supply chain programs that are designed to be flexible, recognizing that many food operations have many different suppliers both domestically and foreign. The rule mandates that a manufacturing/processing facility have a risk-based supply chain program for those raw materials and other ingredients for which it has identified a hazard requiring a supply-chain applied control. Manufacturing/processing facilities that control a hazard using preventive controls, or who follow applicable requirements when relying on a customer to controls hazards, do not need to have a supply-chain program for that hazard. Covered food facilities are responsible for ensuring that these foods are received only from approved suppliers, or on a temporary basis from unapproved suppliers whose materials are subject to verification activities before being accepted for use. If a facility identifies a hazard that they will not control because the hazard will be controlled by a subsequent entity, such as a customer or other processor, the facility will have to disclose that the food is "not processed to control (identified hazard)" and obtain written assurance from its customer regarding certain actions the customer agrees to take.

Part 117.410 General requirements applicable to a supply-chain program:

The supply-chain program must include:

1) Using approved suppliers;
2) Determining appropriate supplier verification activities (including determining the frequency of conducting the activity);
3) Conducting supplier verification activities;
4) Documenting supplier verification activities; and
5) When applicable, verifying a supply-chain-applied control applied by an entity other than the receiving facility's supplier and documenting that verification, or obtaining documentation of an appropriate verification activity from another entity, reviewing and assessing that documentation, and documenting the review and assessment.

The following are appropriate supplier verification activities for raw materials and other ingredients:

1) Onsite audits;
2) Sampling and testing of the raw material or other ingredient;
3) Review of the supplier's relevant food safety records; and
4) Other appropriate supplier verification activities based on supplier performance and the risk associated with the raw material or other ingredient.

Note: Additional regulations may be applicable, see final rules.

The following table summarizes the major differences between a Food Safety Plan vs. a HACCP Plan for human foods:

Food Safety Plan	HACCP Plan
Hazard Analysis	Hazard Analysis
Preventive Controls	CCPs
Parameters, Values and Critical Limits	Critical Limits
Monitoring	Monitoring
Corrective Actions	Corrective Actions
Verification	Verification
Records	Records
Recall Plan	

See also (Fed. Reg., 2015): Table 29, p. 56024.

References

1 Federal Register. 2015. Current Good Manufacturing Practice, Hazard Analysis, and Risk-Based Preventive Controls for Human Food. Vol. 80, Sept 17, 55908.
2 NACMCF. 1997. HACCP Principles and Application Guidelines, August 14, 1997 (see Appendix B).

2

Why Is a Food Safety System the Best Path to Food Safety?

FDA's final FSMA regulations are focused on a preventive approach to food safety. Preventive controls, when properly established and implemented, provide a superior approach over a reactive program, which just fixes and repairs problems after they occur. Preventive controls are also superior to the practice of substantial amounts of end-product testing. It is statistically very difficult to find low levels of contamination in food. To do enough end-product testing to be useful, it would be very expensive and would likely destroy large amounts of product looking for the needle in the haystack. To be truly effective, the preventive control approach must have all elements of the system implemented and operational to be working properly. The system uses a foundation of good manufacturing practices and risk-based preventive controls to mitigate food safety risk. Constructing a working Food Safety System brings together management, line-workers, food safety specialists to design, develop, implement and manage a viable and dynamic system to produce safe and wholesome food.

2.1 What are Biological Hazards and Their Controls?

Each year in the U.S., the Centers for Disease Control and Prevention (CDC) estimates there are approximately 48 million illnesses with 128,000 hospitalizations and 3,000 deaths associated with foodborne illness (CDC, 2011a). Hoffmann et al. (2012) estimated the cost of illness for five of the most common bacterial foodborne pathogens (*Escherichia coli* O157:H7, other Shiga-toxin producing *E. coli*, *Campylobacter*, *Listeria monocytogenes*, and *Salmonella*) as $7.9 billion per year. Although there are limitations to using data estimates, there is no doubt that foodborne illness is a serious problem that warrants attention.

Foodborne illness can result when biological hazards in foods are not properly controlled (CDC, 2011b). An understanding of the types of biological hazards in a specific food is important for an adequate hazard analysis and

FSMA and Food Safety Systems: Understanding and Implementing the Rules,
First Edition. Jeffrey T. Barach.

assignment of effective preventive controls. The characteristics of the microorganism(s) of public health significance must be examined in the context of the food and the environment to determine the appropriate controls.

Surveys have shown that foods most frequently involved in outbreaks are foods of animal origin, although outbreaks from contaminated fruits and vegetables have become more common in recent years, increasing from 0.7% in the 1970s to 6% in the 1990s (Sivapalasingam et al., 2004). Yet when examining the total number of illness, poultry, beef, pork, seafood, and eggs are still most commonly involved (Painter et al., 2013; CDC, 2013).

Ready-to-eat and ready-to-cook foods have also become a new vehicle for outbreaks. Spinach and peanut products were involved in large outbreaks of foodborne illness in 2006 and 2009, respectively. The 2006 outbreak was linked to fresh spinach contaminated with the *E. coli* O157:H7. At least 199 people had been infected, including 3 people who died and 31 who suffered a type of kidney failure called hemolytic uremic syndrome (HUS) (CDC, 2010). Following the 2006 outbreak, various state and federal legislative proposals have emerged that require stricter food production, processing and handling of the food. Industry participants have also adopted improved programs to address food safety. The California Leafy Greens Handler Marketing Agreement (LGMA) developed best practices for farmers that produce lettuce, spinach, and other leafy greens. In May 2011, FDA, USDA, and Cornell University announced the formation of a Produce Safety Alliance to provide produce growers and packers with access to food safety educational materials, science-based best food safety practices, and information about future regulatory requirements.

In late 2008 and early 2009, at least 691 people contracted Salmonellosis from eating products containing peanuts, nine of whom died and 23% were hospitalized (CDC, 2009). FDA confirmed that the outbreak of illnesses was caused by *Salmonella typhimurium* and the source of illness was low-moisture foods including peanut butter, peanut paste, and peanut meal products. Even though *Salmonella* does not grow in low-moisture foods, it can survive and remain viable for extended periods of time. In response to *Salmonella* outbreaks in low-moisture foods, the Grocery Manufacturers Association (GMA) developed several industry guidance documents for controlling *Salmonella* in low-moisture foods (Chen et al., 2009a; 2009b; Scott et al., 2009).

In the past, the development of classic Hazard Analysis Critical Control Points (HACCP) programs focused mainly on the control of microbial hazards. Pathogens in foods remain a top problem for the food industry and a significant public health issue, considering that one in six people are estimated to be stricken with a foodborne illness. The new rules focus on prevention and implementing effective preventive controls to reverse advancing trends in foodborne illnesses.

This guide book will not provide a full review of all possible biological hazards. Other resources are readily available for details about the growth and toxin characteristics, disease properties, and potential food sources of biological hazards in foods. The table below lists many of the common organisms to consider; however, take note that this list is not comprehensive and is for illustrative purposes only. The hazard analysis for identifying potential biological hazards must include a person(s) knowledgeable in bacterial pathogens (vegetative cells and spore formers), toxins from microorganisms, viruses and parasites.

Biological Organism	Example	Potential Hazardous Condition
Bacterial Pathogens		
Vegetative cells	*Eschericia coli (STEC)*............	Fecal contamination of raw foods and ingredients
	Listeria monocytogenes..........	Raw foods, ingredients, and soil, can grow at refrigeration temperatures
	Salmonella spp.......................	Raw foods and ingredients, very durable, dry and heat resistant in low-moisture foods
	Shigella spp............................	Fecal contamination, water or workers
	Staphyloccus aureus..............	Contamination from workers, temperature abuse (toxin)
	Streptococcus group A...........	Contamination from workers handling food
	Vibrio spp................................	Marine seafoods- contamination and temperature abuse
	Yersinia enterocolitica.............	Contamination from raw meats, especially RTE foods.
Spore formers	*Bacillus cereus*.......................	Temperature-abused rice, starchy foods (toxin)
	Clostridium botulinum............	Outgrowth in low-acid foods under anaerobic conditions (toxin)
	Clostridium perfringens...........	Temperature abused meats, stews (toxin)

(*Continued*)

Biological Organism	Example	Potential Hazardous Condition
Viruses	Norovirus................................	Contamination from infected workers handling food
	Hepatitus A.............................	Contamination from infected workers handling food
Parasites	*Campylobacter jejuni*...............	Contamination from raw food and undercooking
	Cryptosporidium parvum..........	Contamination from water and unpasteurized food
	Toxoplasma gondii..................	Contamination from raw meat and undercooking
	Trichinella................................	Contamination from raw meat and undercooking

STEC (Shiga-toxin producing *E. coli*).

In addition to the information provided here, it may be helpful to review more information about specific microorganisms in foods by consulting references, especially the International Commission on Microbiological Specifications for Foods (ICMSF) texts (ICMSF, 2005; 2011) and FDA's Bad Bug Book (FDA, 2013). FDA is developing a Hazards and Control Guide and when this becomes available, it will help industry in hazard assessments of biological hazards (see guidance at: http://www.fda.gov/downloads/Food/GuidanceRegulation/GuidanceDocumentsRegulatoryInformation/UCM517610.pdf).

Classically, there are three major ways of preventing foodborne disease: prevent contamination of the foods, destroy foodborne disease agents that may be present in food, and prevent foodborne disease agents from growing in foods.

Common preventive controls for biological hazards include the following:

- Specifications for microbiological levels in raw materials or ingredients
- Time/temperature controls (thawing/tempering, cooking, freezing, holding, cooling rates, refrigerating, storing, etc.)
- Preservative factors for the food (pH, a_w, etc.)
- Prevention of cross-contamination (e.g., zoning)
- Equipment/environmental sanitation
- Food-handling practices
- Employee hygiene
- Packaging integrity
- Storage, distribution display practices
- Consumer directions for use (to prevent abuse)

2.2 What are Chemical Hazards, Including Allergens and Radioactivity, and Their Controls?

Biological hazards are generally of greatest concern because they are capable of causing widespread foodborne illnesses; however, chemical hazards also have been associated with foodborne illness or injury, albeit generally affecting fewer individuals. Chemical hazards, specifically due to unintended allergens in foods, are a major cause of recalls today. Radiological hazards in foods would be extremely rare but must be considered when developing the Food Safety Plan or during a reanalysis of the hazards and the plan (e.g., purchasing products from an area having a major radiological event like at Fukushima, Japan may warrant a radiological preventive control).

The hazard analysis must consider potential chemical hazards and if present, develop a Food Safety Plan with appropriate control measures. The discussions during the hazard analysis will help decide whether potential chemical hazards warrant inclusion within a Food Safety Plan or whether these potential hazards should be managed within a prerequisite program. This review of chemical hazards, as with other sections on hazards, presents information needed for the identification of *potential* hazards during the first stage of the hazard analysis.

For food production, a wide variety of chemicals are routinely used in the manufacture of foods. The use of some chemicals, such as agricultural pesticides and growth regulators, may not be under the direct control of the establishment. In contrast, some chemicals such as local pesticides, lubricants, sanitizers, and additives for treating water used in processing may be present during production or used throughout the facility or on the facility grounds. Other chemicals may be present or used specifically for particular processes; for example, food colorings solutions may be used in a beverage or dairy formulation. While these chemicals do not present significant hazards when used properly, some of them are capable of causing severe health effects if misused. During the hazard analysis the food safety team and the Preventive Control Qualified Individual must determine if any of these chemicals is reasonably likely to be used in a manner that will result in illness or injury to consumers.

Potential chemical hazards could come from many sources. Here are some to consider:

Chemicals Used as Ingredients or in Formulation	Antimicrobials
	Food Additives (e.g., colors)
	Preservative Compounds
	Nutritional Additives and Vitamins

Chemicals Not Intentionally Added or Are Incidentally Present	Cleaning Chemicals Drug Residues Heavy Metals Industrial Chemicals (not for food use) Pesticides Mycotoxins Radioactive Materials
Allergens, When Unintentionally Present	Supplier Ingredient Is Contaminated Improper Cleaning Cross-contact Improper Labeling
Tampering or Economically Motivated Adulteration	If agents in the adulterated product can cause illness or injury

Food allergies affect an estimated seven to eight million consumers in the U.S. Although most food allergies cause relatively minor or mild symptoms, some can cause very severe reactions that may even be life threatening. Currently, there is no known cure for food allergies. The only successful defense for sensitive consumers is for them to avoid foods containing the causative allergenic proteins.

There are two essential roles for food manufacturers to assist consumers in their efforts to avoid consuming allergenic foods. It is important to have 1) proper labeling of foods, and 2) it is necessary to ensure that foods do not become contaminated with allergenic ingredients due to inadequate cleaning or by cross-contact with allergenic foods that are not declared on a product's label. Nine food types are identified as having proteins that can cause allergic reactions for those individuals having sensitivities. They are:

Foods that have Allergenic Properties
1) Peanut 2) Crustaceans (Shellfish) 3) Wheat 4) Milk 5) Soy 6) Tree Nuts 7) Fish 8) Eggs 9) Other (for products with international distribution, e.g., celery, sesame)

Allergens are considered chemical hazards when cross-contact between non–allergen-containing foods/ingredients and allergen-containing foods/ingredients occurs or when labeling fails to disclose the presence of an allergen. Since allergens affect a small portion of the population, those not afflicted by food allergens may not appreciate the severity experienced by those allergic individuals. It is especially important to educate and train those line-workers involved in plant operations as to the importance of allergen control.

The control of potential chemical hazards in a food-processing facility can be facilitated by the following steps:

- Employ a thorough and comprehensive allergen-management program that addresses storage, cross-contact, cleaning, and label controls
- Use only approved chemicals and develop appropriate specifications. Obtain letters of guarantee from all suppliers of chemicals, ingredients, and packaging materials
- Keep an inventory and post Material Safety Data Sheets of all potentially hazardous chemicals, including food additives and coloring agents, that are used in the establishment
- Review product formulations and current procedures for receiving, storing, and using all potentially hazardous chemicals, as well as procedures for inspecting vehicles for shipping finished products
- Monitor the use of all potentially hazardous chemicals, including the direct observation of employee practices
- Ensure adequate employee chemical safety training, especially for allergens
- Monitor ingredients from suppliers for the potential of economically motivated hazards due to intentional adulteration
- Keep abreast of new regulations and information on allergens and the toxicity of chemicals

2.3 What are Physical Hazards and Their Controls?

Fortunately, many of the foreign materials that end up in food, while aesthetically undesirable, are not considered to be physical hazards. Unlike biological hazards, physical hazards usually create problems only for an individual consumer or relatively few consumers. Physical hazards typically are hard and/or sharp objects that can result in personal injuries such as a broken tooth; lacerations of the mouth, tongue, throat or intestines; or cause choking hazard. Consideration, therefore, must be given to potential physical hazards that can cause injury and to their preventive controls when developing a Food Safety Plan. There are no specific regulations for hard and sharp objects; however, FDA has addressed the issue of adulteration in a guidance document (FDA, 1999). Quality issues, such as low levels of insect parts are important, but do not present a hazard and do not belong in the Food Safety Plan.

Past surveys by the Agency of consumer complaints, have shown hard and sharp objects that caused physical injury (from most to least common) were glass, metal, plastic, stones, shells/pits, and wood. Extraneous matter by regulatory definition also includes such materials as bone fragments in meat and poultry, mold, insects and insect fragments, rodent and other mammalian hairs, sand, and other usually nonhazardous materials. Whether or not these potential physical hazards are controlled in the Food Safety Plan will depend upon an evaluation of the actual likelihood of occurrence and severity of the hazard as determined during the hazard analysis.

As with biological and chemical hazards, there are numerous sources of physical hazards. Potential physical hazards in finished products may arise from sources such as:

- Contaminated raw materials (wood, stones, glass, plastic, metal, shells, pits, etc.)
- Poorly designed or poorly maintained facilities and equipment
- Equipment fragments from wear during operation
- Faulty procedures during production
- Improper employee practices
- Plastic from food containers
- Glass fragments or shards from breakage
- Improper storage of nonproduct materials in food containers
- Tampering or economically motivated adulteration (if injurious)

Prevention and control of potential physical hazards at a facility may include the following:

- Complying with current good manufacturing practices (cGMP) regulations
- Installing equipment that can detect and/or remove potential physical hazards
- Identifying types and sources of potential physical hazards and including controls for hazards requiring a preventive control in the Food Safety Plan
- Sanitary design and preventative maintenance for the design of processing equipment
- Using appropriate specifications for ingredients and supplies
- Obtaining letters of guarantee from all suppliers
- Utilizing vendor certification
- Monitoring consumer complaints for problems and trends
- Training employees

References

CDC, 2009, *Surveillance for foodborne disease outbreaks – United* States, 2006, *Morbid. Mortal. Weekly Rep.* 58, 609–615, viewed 12 July 2016, from http://www.cdc.gov/mmwr/preview/mmwrhtml/mm5822a1.htm.

CDC, 2010, *Surveillance for foodborne disease outbreaks – United* States, 2007, *Morbid. Mortal. Weekly Rep.* 59, 973–979, viewed 12 July 2016, from http://www.cdc.gov/mmwr/preview/mmwrhtml/mm5931a1.htm?s_cid=mm5931a1_w.

CDC, 2011a, *CDC estimates of foodborne illness in the United States,* viewed 12 July 2016, from http://www.cdc.gov/foodborneburden/2011-foodborne-estimates.html.

CDC, 2011b, *Vital signs: Incidence and trends of infection with pathogens transmitted commonly through food – foodborne diseases active surveillance network, 10 U.S. sites, 1996—2010,* viewed 12 July 2016, from http://www.cdc.gov/mmwr/preview/mmwrhtml/mm6022a5.htm?s_cid=mm6022a5_w.

CDC, 2013, *Surveillance for foodborne disease outbreaks – United States, 2009–2010, Morbid. Mortal. Weekly Rep,* viewed 12 July 2016, 62: 41–47, from http://www.cdc.gov/mmwr/preview/mmwrhtml/mm6203a1.htm?s_cid=mm6203a1_w.

Chen, Y., Scott, V. N., Freier, T. A., Kuehm, J., Moorman, M., Meyer, J., Morille-Hinds, T., Post, L., Smoot, L. A., Hood, S., Shebuski, J. & Banks, J., 2009a, "Control of *Salmonella* in low-moisture foods II: Hygiene practices to minimize *Salmonella* contamination and growth," *Food Prot. Trends* 29(7), 435–445.

Chen, Y., Scott, V. N., Freier, T. A., Kuehm, J., Moorman, M., Meyer, J., Morille-Hinds, T., Post, L., Smoot, L. A., Hood, S., Shebuski, J., & Banks, J.. 2009b, "Control of *Salmonella* in low-moisture foods III: Process validation and environmental monitoring," *Food Prot. Trends.* 26(8), 493–508.

FDA, 1999, *Section 555.425: Foods – Adulteration involving hard or sharp foreign objects,* Compliance Policy Guides Manual, Food and Drug Administration, Washington, DC, from http://www.fda.gov/iceci/compliancemanuals/compliancepolicyguidancemanual/ucm074554.htm.

FDA, 2013, *Bad bug book: Foodborne pathogenic microorganisms and natural toxins handbook,* 2nd edn., viewed 12 July 2016, from http://www.fda.gov/Food/FoodborneIllnessContaminants/CausesOfIllnessBadBugBook/.

Hoffmann, S., Batz, M. B. & Morris Jr., J. G., 2012, "Annual cost of illness and quality-adjusted life year losses in the United States due to 14 foodborne pathogens," *J. Food Protect* 75, 1292–1302.

ICMSF, 2005, *Microorganisms in foods 6 – Microbial ecology of food commodities,* 2nd edn., The International Commission on Microbiological Specifications for Foods., Kluwer Academic / Plenum Publishers, New York.

ICMSF, 2011, *Microorganisms in foods 8 – Use of data for assessing process control and product,* The International Commission on Microbiological Specifications for Foods, Springer Science + Business Media, LLC, New York.

Painter, J. A., Hoekstra, R. M., Ayres, T. & Tauxe, R. V., 2013, "Attribution of foodborne illness, hospitalization, and deaths to food commodities by using outbreak data, United States, 1998–2008," *Emerg, Infect. Dis.* 19(3), 407–415.

Scott, V. N., Chen, Y., Freier, T. A., Kuehm, J., Moorman, M., Meyer, J., Morille-Hinds, T., Post, L., Smoot, L. A., Hood, S., Shebuski, J. & Banks, J., 2009, "Control of Salmonella in low-moisture foods I: Minimizing entry of Salmonella into a processing facility, *Food Prot. Trends* 29(6), 342–353.

Sivapalasingam, S., Friedman, C. R., Cohen, L. & Tauxe, R. V., 2004, "Fresh produce: A growing cause of outbreaks of foodborne illness in the United States, 1973 through 1997," *J. Food Protect.* 67, 2342–2353.

3

What are the Essential Elements of a Food Safety System?

The concept of a food safety system approach is superior to both reactive programs to fix problems and to end-product testing. Preventive controls–based systems were embraced in the Hazard Analysis Critical Control Points (HACCP) and are further expanded in the rules of the Food Safety Modernization Act of 2011 (FSMA). To be truly effective, however, the preventive controls approach must have all elements of the system implemented and operational to be working properly. A Food Safety System is defined as:

> A system that a facility implements according to the Food Safety Plan to meet its food safety needs.

In building a Food Safety System, several elements are essential. It starts with a solid foundation of programs that manage the basic environment where safe and wholesome food can be produced. These programs set the stage for the facility and its workers to be able to practice Good Manufacturing Practices (GMPs). Without these GMPs in place, the facility would not be able to have a safe and wholesome operating environment and as a consequence, the high failure rate of preventive controls would make them ineffective. GMP principles and practices are implemented through a series of prerequisite programs, such as cleaning and sanitation programs. These programs are managed through the use of written standard operating procedures (SOPs). Some organizations call these OPRPs or operational prerequisite programs.

The food safety system is focused on the preventive controls that were identified through the hazard analysis as being essential to maintain food safety of the product. These preventive controls (for allergen hazards, sanitation-related hazards, process hazards, and hazards related to suppliers) are carefully designed to manage the hazard to – prevent it, eliminate it, or significantly minimize it to a level of insignificance. The controls are managed through

FSMA and Food Safety Systems: Understanding and Implementing the Rules,
First Edition. Jeffrey T. Barach.

monitoring, and verified through audits, validation, and records. If deviations occur, corrective actions are implemented. The system relies heavily on establishment and maintenance of records so that management and regulatory agencies can see what was done, when, and by whom.

The preventive controls and the operations implementing them are described in the Food Safety Plan. A plan includes a comprehensive hazard analysis and appropriately designed preventive controls. This work and plan is assembled by a Preventive Controls Qualified Individual (PCQI; see Glossary. The Food Safety Plan is the playbook for the facility to make sure the system is working properly. Management has a role here to set the proper food safety culture of the facility and to oversee the implementation and operation of both the Food safety Plan and the Food Safety System.

3.1 What are Prerequisite Programs, and What Do They Do?

The phrase "prerequisite programs" (PRPs) is accepted as an appropriate term to describe a range of programs that are necessary to set the stage for preventive control–based food safety systems and to provide ongoing support for these systems. When properly designed and utilized, prerequisite programs will keep many situations from becoming serious problems that could eventually have an impact on food safety. PRPs provide operating conditions important to the implementation of the Food Safety Plan. Prerequisite programs are an essential component of an establishment's operations and are intended to keep low-risk potential hazards from being likely to occur or becoming serious enough to adversely impact the safety of foods produced.

In the U.S., many of the PRPs are based upon the Food and Drug Administration (FDA) requirements outlined in current Good Manufacturing Practices (cGMPs) (21 CFR part 117 subpart B). In addition to those related to the cGMPs, prerequisite programs can include other systems operations and practices such as ingredient specifications, consumer complaint management, chemical management programs, microbiological monitoring of the plant environment, traceability programs, and supplier approval programs. Without these programs in place and performing effectively, the food safety system may be ineffective in assuring the production of safe foods. Prerequisite programs represent the basic foundation upon which all food safety programs, like HACCP and Hazard Analysis and Risk-Based Preventive Controls are built upon. PRPs are generally managed through the use of standard operation procedure (SOP) written documents (see 4.2). It is important to have adequate training so these SOPs are followed and records are kept properly. With adequate prerequisite programs in place, the development and maintenance of the Food Safety Plan is simplified because many potential hazards are addressed. Although *PRPs generally are not included in the Food Safety Plan,* their maintenance and operation are an important part of the Food Safety System, and those activities have a direct and substantial impact on food safety. *PRPs are not intended to control hazards,* although they may contribute to a potential hazard not becoming a significant hazard. The control of a hazard is assigned to a specific preventive control following the hazard analysis exercise.

Some examples of PRPs follow:

- Facilities designed and maintained to meet cGMP regulations
- Continuing supplier guarantee program and review of supplier's food safety programs
- Written specifications for all ingredients and packaging materials
- Antibiotic residue screening program
- Specifications for equipment to be constructed, installed and cleaned, and sanitized according to sanitary design procedures and principles
- Established and documented preventive maintenance and calibration programs
- A Master Sanitation Program with written sanitation standard operating procedures (SOPs) is highly desirable
- All employees and other persons who enter the manufacturing plant should follow the requirements for personal hygiene
- Documented employee training programs in personal hygiene and plant operations
- Pest control program
- Procedures for proper receiving, storing, and shipping of incoming materials and finished products
- Effective product coding, traceability, and data management systems
- Plantwide temperature and air control SOPs

3.2 What is a Hazard Analysis, and Why is it Performed?

Many in the food industry may not be familiar with what a hazard analysis is or how to go about conducting one. Those regulated by Hazard Analysis Critical Control Points (HACCP) (juice, seafood and meat & poultry products) have experience, whereas those regulated only by GMPs probably have none. Performing one is described in Section 5.6 and can be facilitated using the Hazard Analysis worksheets on pages 71 & 72. FSMA is said to utilize "risk-based" preventive controls to mitigate foodborne hazards. It is appropriate to describe these terms, *risk* and *hazard,* in relation to how they are used in building a food safety system.

First, *risk* is generally an objective term used to describe in quantitative terms an estimate of the size or quality of a hazard. It is generally a mathematical calculation, such as: toxicity (harm) × exposure = risk. It is common to hear that someone has done a risk assessment and the risk is determined to be say, one in one million. In many circumstances, that low level of risk may be acceptable, as it is minimal in relation to other risks. Risks are often considered relative to other risks and to exposure time (e.g., lifetime) to achieve a judgment as to their potential harm. Another example of how risk is used is by CDC. They have estimated the risk of contracting foodborne disease at one in six individuals. Hearing that, we recognize that foodborne disease in the United States is a serious issue and that illness will likely affect someone we know, or perhaps even ourselves. The risk of foodborne disease is something the food industry must address in food service and in food production environments.

A *risk assessment* is described in food safety terms as an opinion or judgment of a given situation, based on the inputs about the harm of the hazard and the exposure to the individual. Risk is used in FSMA then as a judgment about a potential hazard. A potential hazard that has a minimal risk may not need be considered in the list of potential hazards, whereas a known or foreseeable hazard would need to be included. For example, pesticides are used in production of fruits and vegetables. They can be toxic at high levels and cause harm to humans. However, their safe use and further washing fruits and vegetables allows them, under typical circumstances, to be considered "*de minimis*" or of minimal risk and therefore not be put on the list of potential hazards to be evaluated. Following the language of the Food Safety Modernization Act of 2011(FSMA), "risk-based" can then refer to whether a potential hazard gets included in the hazard analysis exercise for further consideration. In the discussions about potential hazards, the PCQI will use judgment to include or exclude potential hazards based on an assessment of their risk.

Those potential hazards that are included in the hazard analysis can then be further evaluated as to their need to be controlled. The need for preventive control is based on the nature of the potential hazard and the particular operations, processes, and procedures for that facility. The hazard analysis is a review of each potential hazard's ability to cause harm. In this review of potential hazards it could be concluded that certain prerequisite programs can adequately mitigate the potential hazard, so it does not need a preventive control applied. The conclusion of the hazard analysis for each potential hazard is a binary outcome (need to control? yes or no). Through knowledge and experience, the PCQI concludes that "the hazard does or does not need an assigned preventive control." This conclusion is based on many factors, both intrinsic and extrinsic, and the rational for the PCQI's conclusion should be recorded in the written hazard analysis. On the Hazard Analysis worksheet (pages 71 & 72), the outcome and rationale is recorded in column #4. The outcome of the hazard analysis should be to address and include potential hazards for that operation by identifying those hazards that are "foreseeable" and "reasonably likely to occur" and therefore require a preventive control. In FSMA, this is described as:

Hazard requiring a preventive control means a known or reasonably foreseeable hazard for which a person knowledgeable about the safe manufacturing, processing, packing, or holding of food would, based on the outcome of a hazard analysis (which includes an assessment of the severity of the illness or injury if the hazard were to occur and the probability that the hazard will occur in the absence of preventive controls), establish one or more preventive controls to significantly minimize or prevent the hazard in a food and components to manage those controls (such as monitoring, corrections or corrective actions, verification, and records) as appropriate to the food, the facility, and the nature of the preventive control and its role in the facility's food safety system.

3.3 What are Risk-Based Preventive Controls, and How are they Assigned?

The Hazard Analysis drives the identification of which potential hazards are significant enough to warrant applying a preventive control to ensure food safety is achieved. Many PCQIs would use a form or a decision approach similar to that described on the form to develop conclusions about hazards and the need for preventive controls. The Hazard Analysis form (see pages 71 & 72) tracks both the ingredients (Part 1) and the steps (Part 2) as described in the process flow diagram. Although this can all be done on a single form, as is often the case for HACCP, many in the food industry find it helpful to use this dual approach of keeping ingredients and process steps separate. This method of hazard analysis is especially useful now that the allergen-containing ingredients may be the basis for assignment of an allergen control.

As mentioned above, the first step is to consider the level of risk from any potential biological, chemical, or physical hazard associated with each ingredient or step. If the risk of that potential hazard at that step or for that ingredient is nonexistent or low enough that it need not be considered, it stays off the list of potential hazards. However, if there is a perceived risk from the potential hazard, it goes on the list and needs to be evaluated for its significance and likelihood of occurring.

Preventive Controls: From Part 117.135

You must identify and implement preventive controls to provide assurances that any hazards requiring a preventive control will be significantly minimized or prevented and the food manufactured, processed, packed, or held by your facility will not be adulterated under section 402 of the Federal Food, Drug, and Cosmetic Act or misbranded under section 403(w) of the Federal Food, Drug, and Cosmetic Act.

Preventive controls include Process Controls, including controls at critical control points (CCPs) and controls that include procedures, practices, and processes to ensure the control of parameters during operations such as heat processing, acidifying, irradiating and refrigerating foods.

Food Allergen Controls include procedures, practices, and processes to control food allergens ensuring protection of food from allergen cross-contact, including during storage, handling, and use; and labeling the finished food.

Sanitation Controls include procedures, practices, and processes to ensure that the facility is maintained in a sanitary condition adequate to significantly minimize or prevent hazards such as environmental pathogens, biological hazards due to employee handling, and food allergen hazards.

Supply-chain controls include the supply-chain program as required by subpart G.

Recall plan as required by § 117.139.

Other Preventive controls could include any other procedures, practices, and processes necessary to satisfy the requirements of controlling hazards. Examples of other controls include hygiene training and other current good manufacturing practices.

Note: Additional regulations may be applicable, see final rules.

3.3.1 What Controls are Used to Control Allergen-Related Hazards?

If allergen hazards are identified in the Hazard Analysis, they may be envisioned at several stages of operation before, during, and after food production. The hazard analysis will direct the need for implementing preventive controls for any or all of these. The following list is not all-inclusive, as each operation is different and suppliers have differences. The choice of preventive controls depends on many factors. Here are some situations where the need for allergen control(s) may be necessary. Some allergen situations to consider are:

- Suppliers' food safety actions for allergen control
- The nature of the ingredients
- The storage area for ingredients
- The steps used to clean equipment
- The separation of areas to avoid cross-contact (zoning)

- The use of dedicated equipment
- Lock-out features of equipment
- Receiving of labels, storage of labels
- Discarding outdated label stock
- Installing the proper label for the product being made
- Any possible other factors depending on the facility and its products

After reviewing the flow diagram and conducting the hazard analysis, it may become apparent that an allergen hazard(s) will exist if proper allergen preventive control(s) are not implemented. These control(s) become part of the Food Safety Plan and may be also part of a more comprehensive Allergen Control Program, if the facility decides to have a general allergen control program.

Here are three typical allergen preventive controls examples. These are only simplified examples, and each facility may require more or less controls, depending on the products and operations.

Ingredient or Step	Hazard	Preventive Control*
Incoming ingredients	Cross-contact with or contains undisclosed allergens; improper storage	Approved supplier; letter of guarantee; audit of supplier; ingredient specification and certificate of analysis
Cleaning equipment	Blind spots where allergens can cross-contact; improper cleaning; ineffective cleaning	SOP for cleaning with routine inspection, especially for food contact surfaces
Labeling product	Label does not have correct allergen information; label mix-up in storage; wrong label on product	Checking labels at receiving; control of labels in storage; discard old labels; check label on product

* This is not a comprehensive list or description, these are only shown as examples.

3.3.2 What Controls are Used to Control Sanitation-Related Hazards?

The high level of attention for initiating sanitation-related controls in FSMA was likely connected to the peanut butter *Salmonella* outbreaks that occurred in late 2008 and early 2009. As a result of the *Salmonella* contamination event, 9 people died, and at least 714 people fell ill from food containing contaminated peanuts. This contamination triggered the most extensive food recall in U.S. history up to that time, involving 46 states, more than 360 companies, and more than 3,900 different products manufactured using peanut butter ingredients. One of the root causes was lack of proper sanitation for a product that is ready to eat. Investigators found evidence of rain and other water leakage into storage areas used for roasted peanuts, practices that allowed for cross-contamination between raw and roasted peanuts, and uncertainty as to whether the peanut roaster routinely reached a temperature sufficient to kill *Salmonella*. With a focus on ready-to-eat products, here are some sanitation situations to consider while doing the hazard analysis:

- Are GMPs implemented and being followed? Is more needed?
- Do raw ingredients have specifications for bacterial loads as confirmed by certificates?
- Does the plant design allow mixing of raw and finished product?
- Does the product flow through production avoid contamination?
- Are workers properly gloved and gowned to prevent product contamination?
- Are ill workers sent home?
- Is product protected from overhead contamination due to moisture, product, filth, etc.?
- Is food contact equipment properly cleaned and sanitized?
- Are drains inspected and tested on a routine basis to minimize contamination?
- Is any rework material used and if so why?
- Is an environmental monitoring program in place and being followed?
- Does packaging equipment protect product during packaging from contamination?

After reviewing the flow diagram and conducting the hazard analysis, it may become apparent that sanitation related hazard(s) will exist if proper sanitation preventive control(s) are not implemented. These control(s) become part of the Food Safety Plan and may be also part of a more comprehensive Sanitation Control Program, if the facility decides to have a general sanitation control program.

Here are four typical sanitation preventive controls examples. These are only simplified examples, and each facility may require more or less controls, depending on the products and operations.

Ingredient or Step	Hazard	Preventive Control*
Incoming ingredients	Contains high bacterial loads and/or pathogens	Approved supplier; letter of guarantee; audit of supplier; ingredient specification and certificate of analysis
Worker product handling	Product contaminated by workers on-the-line	Worker clothing/ protective glove policy; worker illness policy

Ingredient or Step	Hazard	Preventive Control*
Cleaning equipment	Blind spots where bacteria can harborage and cause contamination of product; improper cleaning of food contact surfaces; ineffective cleaning	SOP for cleaning with routine inspection, especially for food contact surfaces
Environmental monitoring and testing	Sampling and testing of equipment and/or production areas to ensure bacterial loads are under control	Environmental monitoring program with strategic testing and actions based on testing results

* This is not a comprehensive list or description, these are only shown as examples.

3.3.3 What Controls are Used to Control Process Hazards?

Process preventive controls can be thought of as the classical critical control points (CCPs) that are described in HACCP systems. In fact, the FDA has suggested that if you have an operational HACCP plan, it makes for an easy transition into a Food Safety Plan to bring into the Food Safety Plan the CCPs that are already working and established. Process controls can be used to control biological, chemical, and/or physical hazards. One significant difference between process controls and other preventive controls is that FDA requires that process controls be validated. For example, if you use a time/temperature heat treatment to inactivate *Salmonella* in a processing step, you will need to have scientific data to show that that specific treatment will sufficiently kill/reduce the pathogen to an acceptable level. This validation report becomes part of the documents of the Food Safety Plan. Process controls, as in HACCP, have specific critical limits, whereas other preventive controls may rely on parameters or other criteria to meet to establish conformance. Some situations in the manufacturing steps to consider that may suggest a process control is needed are:

- Ingredients may contain a variety of possible microbial hazards, viruses, biological hazards, toxins, aflatoxins (mycotoxins), or other microbial or chemical hazards.
- Is potable water and ice used as an ingredient?
- Incoming ingredients may have metal, wood, stones, etc. that need to be removed in the process.
- pH and/or water activity of the product is critical to ensure pathogens won't grow and produce toxin.
- A heating step is used to sufficiently kill/reduce Salmonella (or another pathogen) to an acceptable level.

- Equipment used to control time/temperature or other process parameters should be properly sized, maintained, and able to adequately establish control.
- Glass containers are used, and glass breakage is of concern.
- Product is heated then cooled, and if improperly cooled, pathogens can grow.
- Cutting/shedding or blending equipment have blades that can fragment if misaligned or worn and result in metal fragments.
- A ready-to-eat product relies on refrigeration for food safety during storage and shipping.

Here are four typical process preventive controls examples. These are only simplified examples, and each facility may require more or less controls, depending on the products and operations.

Ingredient or Step	Hazard	Preventive Control*
Incoming ingredients	Contains high bacterial loads and/or pathogens	Use a validated time/temperature heating/cooking process to inactivate the pathogen of concern
Incoming ingredients have metal or wood; metal can come from process equipment wear	Metal and wood fragments are physical hazards that can cause mouth injury	Screens, magnets, and or metal detectors can remove physical hazards
Water activity, pH, salt level, chemical preservatives	Food preservation additives and conditions will not prevent hazards if out of control	Have established critical limits for preservative steps to ensure pathogens don't grow
Thermal treatments generate heated product that may contain bacterial spores	Spores from pathogens (*C. perfringens, B. cereus*) will survive most heating steps and can grow with toxin production if not properly cooled	Cooling of product to critical limits is required to prevent growth and toxin production in cooked food

* This is not a comprehensive list or description, these are only shown as examples.

3.3.4 What Controls are Used to Control Supplier-Related Hazards?

Biological, chemical, and physical hazards could be associated with the incoming ingredients and materials that are purchased from suppliers. If you identify a hazard and your facility does not mitigate the hazard, you either have to have it controlled by your supplier, or if the hazard remains uncontrolled in your product, you must alert purchasers of your product that it may contain an uncontrolled hazard. Supplier controls can be essential to the safety of the product, and when relying on your supplier to control a hazard, you must actively have a program that manages that supplier's control. It starts with having criteria for the approval of suppliers. If a supplier does not meet the criteria, it's best to drop them. Using approved suppliers helps to ensure they are giving attention to a serious food safety issue. There are several steps in approving a supplier, but this will not be addressed here. Companies must work with their procurement staff and supplier to establish these. What FSMA requires is that you verify that the supplier is properly controlling the hazard if its control is described in the specifications for his ingredient. Verification usually takes the form of an audit of his operations; yearly if the hazard is of such a nature as to cause severe harm or death – the type of hazard that would be a Class I recall if not controlled. The audit would be a review of the supplier's Food Safety Plan and facilities as well as review of records and any problems and deviations from conformance to your specifications. If the hazard is not controlled by your supplier or by you, you must notify your customers in writing to that fact so they can address the issue as may be necessary.

Economically motivated adulteration (EMA) is the fraudulent, intentional substitution or addition of a substance in a product for the purpose of increasing the apparent value of the product or reducing the cost of its production, that is, for economic gain. With respect to FSMA rules, it applies to only those agents that can cause illness or injury. When a preventive control is needed, a supply chain program is a typical approach to managing the hazard using audits and testing procedures. Examples of past EMA situations include melamine in milk products to improve apparent quality and protein content; the use of illegal food colors like malachite green in seafood and Sudan Red in chili peppers; and foods contaminated with diethylene glycol, a toxic syrup substitution.

Identifying supplier controls would take place after reviewing the flow diagram and conducting the hazard analysis. It should become apparent that certain hazard(s) are controlled by the supplier and that if proper preventive control(s) are not implemented by the supplier, your product could be at risk. These supplier control(s) become part of both his and your Food Safety Plans. Some example situations where supplier controls may be used are:

- Nuts are used as an ingredient for a ready-to-eat food, and the supplier uses a pasteurization process to inactivate *Salmonella*;
- Lettuce is an ingredient in an uncooked product, and the supplier triple-washes the lettuce to remove soil and bacteria;

- Frozen peas are used in a pasta salad, and blanching is important to food safety;
- Spices are part of an uncooked entrée, and the supplier uses irradiation to eliminate pathogens;
- Milk products from suppliers in certain countries have been shown to have a pattern of economically motivated adulteration; verification of authenticity is important.

Here are three typical supplier preventive controls examples. These are only simplified examples, and each facility may require more or fewer controls, depending on the products and operations.

Ingredient or Step	Hazard	Preventive Control*
Ingredients from approved supplier, treated for pathogens	If not treated properly, may contain high bacterial loads and/or pathogens	Supplier controls hazard by a validated process; purchase specifications describe the control; supplier provides a letter of guarantee that process is done; audit of supplier's conformance
Incoming ingredients, supplier makes ingredients with and without an allergen	Cross-contact could occur in the supplier's facility, and the ingredient you purchase could have an undisclosed allergen	Use an approved supplier to meet your specification of no cross-contact; letter of guarantee or certificate of analysis upon receipt of ingredient
Fresh fruits and vegetables used without further cooking	Fruit and vegetables grown in unsanitary conditions may harbor pathogens and/or chemical contaminants	Farmer supplier uses good agricultural practices (GAPs) to produce and harvest vegetables; letter by farmer he is following FSMA produce safety rules and GAP sanitary produce practices.

* This is not a comprehensive list or description, these are only shown as examples.

3.4 What is a Food Safety Plan, and Who Develops It?

As stated above, the Food Safety Plan is the playbook for the Food Safety System. A Food Safety Plan is defined as:

> A set of written documents that is based upon food safety principles; incorporates hazard analysis, preventative controls, supply-chain programs, and a recall plan; and delineates the procedures to be followed for monitoring, corrective action, and verification.

Those familiar with HACCP will quickly observe that the FSMA Food Safety Plan is basically an extended and more precise HACCP plan. The HACCP plan includes basically the hazard analysis, Process Controls to address the reasonably foreseeable hazards, and some other Critical Control Points (CCPs) for additionally identified foreseeable hazards such as allergens, and so on. The FSMA Food Safety Plan includes the hazard analysis, the four major types of Preventive Controls (Process, Allergens, Sanitation and Supplier Controls) and also includes a Recall Plan.

FSMA does not specify a structure (no specific forms or format) of the Food Safety Plan but does outline the purpose and function of the plan. The Agency recognized that many companies have their own outlines and forms to structure their own plans and keep records, so the concept here is to be flexible but descriptive about what has been done to address hazards and what will be done to prevent their occurrence in foods. The forms used in this book are the author's suggested forms to help the reader understand what content is needed and what actions will fulfill the regulatory requirements. Developers of Food Safety Plans can use these forms, find those developed by others, or develop their own forms and Food Safety Plan formats, as long as they meet the regulatory requirements. As a tool to assist in the Food Safety Plan development, Appendix A is a simplified FSP Checklist.

The hazard analysis step toward developing a plan is generally a complex process and requires knowledge and skill to get it right. The regulatory requirement here is to have a Preventive Controls Qualified Individual (PCQI) oversee conducting the hazard analysis and the development of the Food Safety Plan. A PCQI may or may not be an employee of the company:

Preventive Control Qualified Individual (PCQI)
One who has successfully completed training in the development and application of risk-based preventive controls at least equivalent to that received under a standardized curriculum recognized as adequate by FDA or is otherwise qualified through job experience to develop and apply a food safety system.

4

How is a Food Safety System Managed?

4.1 What is the Role of Management and Plant Operations in a Food Safety System?

Consumer trust is one of the most important aspects of the food business. That trust is built upon a foundation of food safety. As in the past, the responsibility for food safety is shared between the regulatory agencies and industry. With FSMA now in place, that responsibility has made a subtle but important shift. The Food Safety Modernization Act of 2011 (FSMA) is built upon a premise that it explicitly embraces and enhances the food safety role of the food industry (Taylor, 2015). This assertion relies on the core principle that those who produce food for the commercial market have the responsibility and capability to make it safe in accordance with recognized best practices for preventing illness and injuries. In the past, FDA's food safety oversight was mostly at arm's length from the industry, inspecting plants and investigating problems after they had occurred. FDA defines its new role as a job of establishing the framework of standards that help define food safety practices and then overseeing their implementation to achieve high rates of compliance with these standards. This means FDA will be carrying out inspections in a way that focuses less on possible regulatory violations and more on whether food producers are meeting their responsibility to achieve good food safety outcomes (Taylor, 2015). Their oversight role will focus more on verification of what industry is doing to prevent problems, and they will be functioning more as an integral part of the food system, rather than standing apart from the system as primarily an enforcer of rules by inspection.

Management then has an expanded role to establish and maintain consumer trust, produce safe and wholesome food, and to meet regulatory compliance requirements. It begins with establishing a "food safety culture" within the organization. Top management must set the tone and provide resources for

FSMA and Food Safety Systems: Understanding and Implementing the Rules,
First Edition. Jeffrey T. Barach.

this to happen. Management must institute a food safety system that is risk-based and backed up by science. It would include rigorous prerequisite programs that address Good Manufacturing Processes (GMPs) and a Food Safety Plan based on a plant and food product's specific hazard analysis. Appropriate preventive controls would be assigned to hazards and these would be managed by those in charge of plant operations. They would carry out the day-to-day monitoring, record-keeping, verifications, and corrections if necessary. Management also needs a crisis management and recall plan in place if one should become necessary.

To ensure the system will operate as designed, management must provide staff with guidance, education, training, and technical assistance so they know what's expected and that they are supported in doing it correctly. For an effective management of food safety over and above science, well-designed systems, equipment and procedures, and consideration of human factors are all essential. This ranges from factors underlying consumer choice and practices, to commitment and motivation of managers in the food industry providing adequate infrastructure and organizational culture conducive to professional food safety management (Motarjemi and Lelieveld, 2013).

4.2 How are SOPs Developed and Managed?

A Standard Operating Procedure (SOP) is a set of written instructions that document a routine or repetitive activity followed by an organization (USA EPA, 2007). SOPs give a step-by-step description of how a specific operation, method, or procedure is performed. The instructions give details and specify documents for an operational activity, such as that associated with cleaning and sanitizing equipment that comes in contact with food during production. Developing and using SOPs are an integral part of a successful food safety system as it provides individuals with the information to perform their job properly. SOPs for management of current Good Manufacturing Processes (cGMPs) and Food Safety Plan preventive controls ensure consistency in the food safety aspects of the manufacturing environment and the safety of the product.

As may be evident, if SOPs are not written properly or if employees fail to follow them correctly, they are of limited value. The success of SOPs depends upon their being well written, thoroughly reviewed, and re-enforced by management's direct supervisors. SOPs should be reviewed (that is, validated) by one or more individuals with appropriate training and experience with the process. It may be helpful if SOPs are actually tested by individuals other than the original writer before the SOPs are finalized. Current copies of the SOPs also need to be readily accessible for reference in the work areas of those individuals actually performing the activity, either in hard copy or electronic format; otherwise, SOPs that are not readily accessible serve very little purpose.

SOPs should be written with sufficient detail so that someone with limited experience with or knowledge of the procedure can successfully reproduce the procedure when unsupervised. The experience required and subsequent training for performing an activity should be noted in a section on personnel qualifications.

The organization should maintain a master list of all SOPs. This file or database should indicate the SOP number, version number, date of issuance, title, author, status, organizational division, branch, section, and any historical information regarding past versions. SOPs should be organized to ensure ease and efficiency in use and to be specific to the organization that develops it.

FSMA does not require any particular format for SOPs, and there is no one "correct" format. Organizations will develop their own, and they may vary depending on the type of SOP being written. The level of details provided in the SOP may differ quite extensively, depending on whether the process is critical to food safety, considered a routine operation, or fulfills a regulation. Other attributes of a SOP include the frequency of the procedure being followed, the number of people who will use the SOP, and the availability and time required for training.

A generalized format for an SOP outline is shown below (USA USDA, 2005). Consult references for examples and useful templates:

Outline for Developing and Managing a SOP

Title Page: The first page or cover page of each SOP should contain the following information: a title that clearly identifies the activity or procedure, an SOP identification (ID) number, date of issue and/or revision, the name of the applicable agency, division, and/or branch to which this SOP applies, and the signatures and signature dates of those individuals who prepared and approved the SOP.

Table of Contents: A Table of Contents may be needed for quick reference, especially if the SOP is long, for locating information and to denote changes or revisions made only to certain sections of an SOP.

Definitions: Define any specialized or unusual terms either in a separate definition section or in the appropriate discussion section. This section lists definitions of terms, acronyms and abbreviations relevant to this SOP, or with which the reader may be unfamiliar.

Text of Procedures: Provides a step-by-step description of the operation. Well-written SOPs should first briefly describe the purpose of the work or process, including any regulatory information or standards that are appropriate to the SOP process, and the scope to indicate what is covered. Describe what sequential procedures should be followed, divided into significant sections, if needed. Use of diagrams and flow charts may help to break up long sections of text and to briefly summarize a series of steps for the reader.

Monitoring: Describes the steps for real-time routine observations and records that show the procedures are being done correctly.

Corrective Actions: Provides a step-by-step plan to correct any procedures that were carried out incorrectly or that did not accomplish the objective. Disposition of product and root cause of the problem should be considered. Actions may be as simple as repeating the given procedure but may be more complex including revising the procedures and/or retraining staff.

Personnel Qualifications and Training: This section identifies the minimal education or training that is required to carry out the procedure covered by the SOP.

Records and Verification: Name the records developed from the SOP activity and who signs them and where they are sent and archived. Included are any documents that validate the SOP and review of records that verify procedures are correctly accomplished.

Resources and References: Attach any appropriate information, e.g., an SOP may reference other SOPs. This section also lists any document used as a source for writing the SOP such as standard methods, QA Manual, publications, and instrument manuals.

Revising: SOPs are reviewed and revised periodically (e.g., biannually) to ensure that policies and procedures continue to be relevant and accurate. It may be revised prior to the review cycle if a modification or change to procedure is required. Revision may include: revise the SOP text as necessary, include any relevant identification numbers, create a new version if required, submit the new or revised SOP for review and approval, and replace old versions with new.

DATE IMPLEMENTED: ______________ **BY:** ______________________
DATE REVIEWED: __________________ **BY:** ______________________
DATE REVISED: ____________________ **BY:** ______________________

USA. Environmental Protection Agency (EPA) (2007). Guidance for Preparing Standard Operating Procedures (SOPs). EPA QA/G-6. EPA/600/B-07/001.

USA. U.S. Department of Agriculture (USDA), Food and Nutrition Service & National Food Service Management Institute. National Food Service Management Institute (2005). HACCP-Based Standard Operating Procedures (SOPs). University, MS: Author.

4.3 How are Preventive Controls Managed?

Preventive controls, required under FSMA, are subject to the following preventive control management components as needed to ensure the effectiveness of the preventive controls, taking into account the nature of the preventive control and its role in the facility's food safety system:

- Monitoring in accordance with Part 117.145
- Corrective actions and corrections in accordance with part 117.150
- Verification in accordance with Part117.155
- Control management components as appropriate to ensure the effectiveness of the supplier controls (subpart G)
- Review of records in accordance with Part 117.165(a)(4)

These components are further described in the following sections.

4.3.1 What are Performance Criteria for Controls, Including Critical Limits?

For preventive controls to be effective, they need a range of operation established which defines whether a hazard is "being controlled" or "is not being controlled." This range could be a critical limit, which usually has a maximum and/or minimum value. Critical limits typically are used for process preventive controls, and this application is comparable to establishing a critical limit in a Hazard Analysis Critical Control Points (HACCP) system (Principle #3) for control of a Critical Control Point (CCP). A critical limit is used to distinguish between safe and unsafe operating conditions at a CCP.

Critical limits are safety-based and should not be confused with operating limits, which are typically quality-based or established for other reasons. For example, a baking process may kill *Salmonella* at 160 °F instantly but to obtain a brown crust on the baked good, the internal temperature must reach 200 °F in the oven. In that case the operating limit is 40 °F above the food safety limit. Operating limits are not used in the Food Safety Plan, only critical limits. Since a critical limit typically is met or exceeded, critical limits are generally expressed as greater than or less than so as to be flexible (example, use ≥ 160 °F and not just 160 °F or ≤ 41 °F in 4 hours).

Some preventive controls have a binary (yes or no; +/−) type criteria for control. Either the operation was "done" or "not done"; the action was completed "yes" or "no" to define if the hazard is being controlled. A performance criteria example for a sanitation-related preventive control – Was the food contact surface properly cleaned and sanitized according to the procedure (yes or no)? An example for a supplier-related preventive control – Did the receiving clerk collect the certificate of analysis (COA) upon receipt of the raw materialm and did it say "*Salmonella* test: negative" (yes or no)?

Each preventive control will have some way to measure that the identified hazards are prevented, eliminated, or reduced to acceptable levels. That control measure has one or more associated critical limits or some other performance criteria. Critical limits for process controls may be based upon factors such as temperature, time, physical dimensions, humidity, moisture level, water activity (a_w), pH, titratable acidity, salt concentration, available chlorine, viscosity, preservatives, or sensory information such as aroma and visual appearance. Critical limits should be scientifically based. For process controls, this means the control action should have a document(s) showing the limits for the control have been validated (see section 4.3.5) and that the criterion for food safety has been met. An example of a criterion is a specific lethality of a cooking process such as a 5 log reduction (D-value) in *Salmonella*. As described in the validation section below, the critical limits and criteria for food safety may be derived from sources such as regulatory standards and guidelines, literature surveys, experimental results, and experts. For some process controls, the limits may be based on actual measurements of the product (say a thermometer inserted in the baked food to ensure a temperature of the product ≥ 160 °F), or they may be based on an established process using a specific piece of equipment (example: in-house validation data shows oven #2 operating at 350 °F with a belt speed of 1 ft/minute will deliver a process to a baked good giving an internal temperature of ≥ 160 °F).

4.3.2 How are Preventive Controls Monitored?

In order to determine if a preventive control is actually controlling a hazard, and limits are not being exceeded, the preventive control needs to be monitored. Those familiar with HACCP will recognize this activity as the basis for Principle #4.

Monitoring is a planned sequence of observations or measurements to assess whether a preventive control is under control and to produce an accurate record for future use in verification. Monitoring serves three main purposes. First, monitoring is essential to food safety management in that it facilitates tracking of the operation. If monitoring indicates that there is a trend towards loss of control, then action can be taken to bring the process back into control before a deviation from a critical limit occurs. Second, monitoring is used to determine when there is loss of control and a deviation (failure to meet limit criteria) occurs at a preventive control, that is, exceeding or not meeting a critical limit. When a deviation occurs, an appropriate corrective action (see section 4.3.3) must be taken. Third, it provides written documentation for use in verification.

Monitoring procedures must be effective. Ideally, monitoring should be continuous, which is possible with many types of physical and chemical methods. When it is not possible to monitor on a continuous basis, it is necessary to establish a monitoring frequency and procedure that will be reliable enough to indicate that the preventive measure is under control. Statistically designed data collection or sampling systems lend themselves to this purpose. There are many ways to monitor critical limits on a continuous or batch basis and to record the data on charts. Continuous monitoring is always preferred when feasible. Monitoring equipment must be carefully calibrated for accuracy.

In addition to being trained in the monitoring technique for which they are responsible, employees should be trained in procedures to follow when there is a trend towards loss of control so that adjustments or corrections can be made in a timely manner to assure that the hazard remains under control before a deviation occurs. The person responsible for monitoring must also immediately report a process or product that does not meet critical limits.

Examples of monitoring activities include visual observations and measurement of temperature, time, pH, and moisture level, and so on. Microbiological tests are seldom effective for monitoring due to their time-consuming nature and problems with ensuring detection of contaminants.

Monitoring: From Part 117.145

As appropriate to the nature of the preventive control and its role in the facility's food safety system:

You must establish and implement written monitoring procedures as part of the Food Safety Plan, including the frequency with which they are to be performed.

You must monitor the preventive controls with adequate frequency to provide assurance that they are consistently performed.

You must document the monitoring of preventive controls with records that are subject to verification and records review.

Note: Additional regulations may be applicable, see final rules.

4.3.3 If Preventive Controls Fail, What Corrective Actions are Needed?

The aim of the Food Safety System is to identify health hazards and to establish strategies to prevent, eliminate, or reduce their occurrence. However, in the real world ideal circumstances do not always prevail, and deviations from established processes and operations can occur. An important function of corrective actions, whether planned ahead or following an unanticipated problem is to prevent foods that may be hazardous from reaching consumers. Where there is a deviation from established critical limits and parameters, corrective actions are necessary. Therefore, corrective actions should include the following elements:

a) determine and correct the cause of noncompliance
b) determine the disposition of noncompliant product
c) record the corrective actions that have been taken

FSMA requires that specific corrective actions should be developed in advance for each preventive control and included in the Food Safety Plan. As a minimum, the Food Safety Plan should specify what is to be done when a deviation occurs, who is responsible for implementing the corrective actions, and that a record will be developed and maintained of the actions taken. Individuals who have a thorough understanding of the process, product, and Food Safety Plan should be assigned the responsibility for oversight of corrective actions. As appropriate, experts may be consulted to review the information available and to assist in determining disposition of a non-compliant process or product.

Corrective Actions and Corrections: From Part 117.150

As appropriate to the nature of the hazard and the nature of the preventive control:

You must establish and implement written corrective action procedures as part of the Food Safety Plan that must be taken if preventive controls are not properly implemented or the control or the Food Safety Plan is found to be ineffective.

The corrective action procedures must describe the steps to be taken to ensure that:

- Appropriate action is taken to identify and correct a problem that has occurred with implementation of a preventive control
- Appropriate action is taken, when necessary, to reduce the likelihood that the problem will recur
- All affected food is evaluated for safety
- All affected food is prevented from entering into commerce

In the event of an unanticipated food safety problem, and you do not have a written corrective action, you are subject to take corrective actions similar to the requirements as described above.

Exception by Correction: For certain situations where you are able to take action, in a timely manner, to identify and correct conditions with food allergen controls, sanitation-related controls, or to identify and correct a minor and isolated problem that does not directly impact product safety, this is not considered a corrective action.

Records: All corrective actions (and, when appropriate, corrections) taken in accordance with this section must be documented in records. These records are subject to verification and records review.

Note: Additional regulations may be applicable, see final rules.

4.3.4 How is the System and its Parts Verified as Being Compliant?

FSMA describes verification as the application of methods, procedures, tests, and other evaluations, in addition to monitoring, to determine whether a control measure or combination of control measures is or has been operating as intended and to establish the validity of the food safety plan.

In a general sense, verification activities will take place by both the Agency and by the food company. As we described earlier, FDA's role is changed under FSMA to rely less on inspections and more on verification that the company is properly managing its food safety system. This section will focus on the industry's role in verification.

One confusing issue that has repeatedly been discussed in food safety circles is – What is the difference between verification and validation, and why is validation a subset of verification? A Summit was held in 2013 to address this issue, and a framework was established to help industry fit these activities into the requirements of FSMA (Brackett, R. E. et al., 2014). The conclusion of the industry and academic participants was to recognize the importance of both these activities to achieve FSMA goals and that it was best to consider them as two separate tasks. Industry must use science-based tools to construct schemes that validate the effectiveness of preventive controls for hazards. Also, industry management should set a "food safety culture" and manage and verify the food safety system with responsible staff who can recognize and adapt when change is needed. Several worksheets and checklists were developed supporting the framework with multiple examples of verification and validation activities.

An important aspect of verification is the initial verification of the Food Safety Plan to determine that the plan is scientifically and technically sound, that all hazards have been identified, and that if the plan is properly implemented these hazards will be effectively controlled. As a subset of verification, initial validation of process preventive controls is needed to provide data and information that the process controls are science-based. Validation is discussed as a separate task (see section 4.3.5). It should be noted that FSMA only requires process preventive controls to be validated, although a company can validate other preventive controls if it wishes to also ensure these controls are effective.

The process of establishing verification activities is part of the development of the written Food Safety Plan and should take place during its development and implementation. Ongoing verification is also necessary to evaluate the working plan to make sure the system is functioning according to the plan. Verification becomes an integral part of the Food Safety System. It is used to confirm that the plan is designed well and that it is working as intended and that the plan is being correctly followed.

Verification of the system and the Food Safety Plan can be thought of as an audit-type activity. A periodic and comprehensive verification of the Food

Safety System should be conducted by an unbiased, independent authority. Such authorities can be internal or external to the food operation. This should include a technical evaluation of the hazard analysis and each element of the Food Safety Plan as well as on-site review of all flow diagrams and appropriate records from operation of the plan. A comprehensive verification is independent of other verification procedures and must be performed to ensure that the Food Safety Plan is resulting in the control of the hazards. If the results of the comprehensive verification identifies deficiencies, the Food Safety Team must modify the plan as necessary.

Routine verification activities are also carried out by individuals within a company to ensure monitoring activities are done correctly and at the specified frequency.

Verification: From Part117.155
Verification activities must include, as appropriate to the nature of the preventive control and its role in the facility's food safety system: 1) Validation in accordance with Part117.160 [see section 4.3.5] 2) Verification that monitoring is being conducted as required by Part 117.140 (and in accordance with Part 117.145) 3) Verification that appropriate decisions about corrective actions are being made as required by § 117.140 (and in accordance with § 117.150) 4) Verification of implementation and effectiveness in accordance with Part 117.165 5) Reanalysis in accordance with Part 117.170. Documentation: All verification activities conducted in accordance with this section must be documented in records.Note: Additional regulations may be applicable, see final rules.

4.3.5 How are Process Preventive Controls Validated?

FSMA defines validation to mean obtaining and evaluating scientific and technical evidence that a control measure, combination of control measures, or the food safety plan as a whole, when properly implemented, is capable of effectively controlling the identified hazards.

Information needed to validate the overall Food Safety Plan often include (1) expert advice and scientific studies and (2) in-plant observations, measurements, and evaluations. For example, validation of the cooking process for a refrigerated soup product should include the scientific justification of the heating times and temperatures needed to obtain an appropriate destruction of pathogenic microorganisms (e.g., enteric pathogens) and studies to confirm that the conditions of cooking will deliver the required time and temperature to each container of soup. Further validation may include studies or literature showing that cooling is sufficient to prevent any pathogen growth before packaging.

Initial validations are performed when a product process is established. This ideally should occur several months before operations begin. The Preventive Controls Qualified Individual (PCQI) and others on the Food Safety Team (or outside consulting individuals) develop a validation plan, conduct studies, and write a validation report for inclusion in the Food Safety Plan. A reanalysis (revalidation) may be called for if there are unexplained system failures, significant product and/or process changes, throughput changes, significant packaging changes, or if new food safety hazards are recognized. In any case, a reanalysis (Part 117.170) overseen by a PCQI should occur every three years or sooner, if warranted.

The following are steps involved in the validation process and potential reasons for conducting a reanalysis (Brackett, R.E. et al., 2014):

1) Assemble relevant validation information and data, conduct studies where needed for missing information:
   - Evaluation of product and process – Are all product descriptions available and correct food safety characteristics identified, and do all products have a valid process flow?
   - Product category safety, history, and trends – Are there any emerging hazards that need to be included, and are there changes/trends/consumer complaints that indicate process capability needs to be reassessed?
   - Preventive control management – Are the appropriate preventive controls applied, and are the preventive controls documented appropriately at the plant?
2) Analyze the results from the information and data collected for food safety implications:
   - Determine and implement corrective actions
   - Decide if the step, procedure, SOP, program and control measure are sufficient and can be implemented
   - If not, modify parameters, equipment, procedures, etc. to address the control measure
3) Document the results and approve the validation plan:
   - Management agrees with results of the validation plan – with signatures
   - Management is engaged with implementation of the validated plan
   - Documentation identifies conditions when the plan needs to be revalidated
   - Documentation is archived for use during reanalysis/revalidation
4) Potential reasons for validation reanalysis:
   - It has been three years since last validation
   - A new product, process, or piece of processing equipment has been added or changed
   - A significant change in formulation has occurred that suggests a need for reanalysis

- New/different hazard(s) have been identified
- Preventive controls are not appropriate or no longer controlling hazards
- Critical limits or parameters are no longer valid (e.g., due to new experimental/regulatory data)
- Monitoring actions (procedures and/or frequency) are no longer assessing the effectiveness of the preventive control
- Corrective actions are too frequent and/or not effective
- Ongoing verification activities (including validation) do not ensure that food safety system is adequate to control hazards and that the procedures are consistently being followed
- Records do not provide adequate documentation that the procedures/employees are doing what they are assigned to accomplish
- A new or updated regulatory requirement is identified

4.4 How are Records Established and Maintained?

Records are an important component of both past and modern food safety systems and are an increasing regulatory requirement. With the role of food safety squarely on the shoulders of the food industry, FDA relies more and more on records to provide an accurate account of who did what and when. Review of records and documents by FDA inspectors is becoming a bigger part of their job in evaluating the operations at a facility. Therefore, providing accurate and complete records that are available in a timely manner is very important for both the company and the Agency. Failure to keep and provide records can result in regulatory and financial problems. Product may be recalled due to lack of or incomplete records. Records provide proof that the Food Safety System was operating as designed. If records are unavailable, proof does not exist for FDA. It is often said "if it's not written down, it didn't happen."

In addition to FSMA regulations, the Bioterrorism Preparedness and Response Act of 2002 (BT Act) requires facilities that manufacture, process, pack, transport, distribute, receive, hold, or import food to establish and maintain records regarding the previous source of all food and ingredients and the subsequent recipients of all food and ingredients shipped into commerce. This is often referenced as "one step forward and one step back" record-keeping. As well as FSMA regulations requiring record establishment and maintenance, the BT Act incorporates a similar scope of record-keeping activities as identified for FSMA requirements with the addition of the "one step forward and one step back" requirement and having records available on a twenty-four-hour or less basis. The BT Act requires food processors to establish, maintain, and have access to records from their operation so that a product or ingredient that presents a threat of serious adverse health consequences or death can be thoroughly tracked and recovered rapidly. For any food safety system, records also facilitate the activities that take place during a market withdrawal, stock recovery, or product recall.

FSMA introduces the use of exception records to supplement those required in HACCP as well as those required by the BT Act and other regulations. These records document temperature controls for refrigerated foods. Records of refrigeration temperature during storage of food that requires time/temperature control to significantly minimize or prevent the growth of, or toxin production by, pathogens may be affirmative records demonstrating that temperature is controlled or exception records demonstrating loss of temperature control. Exception records may be adequate in circumstances other than monitoring of refrigeration temperature.

Record-keeping for FSMA includes records that go beyond those that are maintained during the day-to-day operation of the Food Safety Plan. FSMA requires records to be established and maintained for a minimum of two years. Some other regulations may require more or less record retention. For example, for low-acid canned foods three years retention is required. A well maintained food safety system also includes generally four types of records supporting the development of the plan and the system:

1) Summary of the hazard analysis
2) The Food Safety Plan
3) Support documentation
4) Daily operational records

Record-keeping: From Parts 117.301, 305, 310 and 315

General requirements applying to records, they must:

- Be kept as original records, true copies (such as photocopies, pictures, scanned copies, microfilm, microfiche, or other accurate reproductions of the original records), or electronic records
- Contain the actual values and observations obtained during monitoring and, as appropriate, during verification activities
- Be accurate, indelible, and legible
- Be created concurrently with performance of the activity documented
- Be as detailed as necessary to provide history of work performed and include:
  1) Information adequate to identify the plant or facility (e.g., the name, and when necessary, the location of the plant or facility)
  2) The date and, when appropriate, the time of the activity documented
  3) The signature or initials of the person performing the activity
  4) Where appropriate, the identity of the product and the lot code, if any

General requirements for record retention:

All records required by this part must be retained at the plant or facility for at least 2 years after the date they were prepared.

Records that a facility relies on to support its status as a qualified facility must be retained at the facility as long as necessary to support the status

Except for the Food Safety Plan, offsite storage of records is permitted if such records can be retrieved and provided onsite within 24 hours of request for official review. The Food Safety Plan must remain onsite. Electronic records are considered to be onsite if they are accessible from an onsite location.

Note: Additional regulations may be applicable, see final rules.

4.5 Why and How is a Recall Plan Developed and Managed?

Recalls are procedures used by the food industry to identify and recover potentially adulterated, misbranded and/or hazardous foods in order to prevent or limit potential food safety problems or economic fraud. The main purpose of a recall is to protect consumer health and comply with existing rules and regulations. These product recovery procedures have been traditionally part of industry practices that have proven to be effective for all foods, in all levels of commerce. FSMA regulations have specified that a facility must have a recall plan as a preventive control measure. The design and scope of that plan is fairly flexible in the regulations, so a facility has some latitude about how the plan is structured, what form the written plan takes, and how the plan is implemented if a recall should be necessary. Before FSMA regulations became effective, FDA relied heavily of industry to do voluntary recalls, with their assistance. FDA still expects industry to conduct voluntary recalls; however, FSMA gives additional authority to FDA to require a company to conduct a mandatory recall if FDA believes a recall is necessary to protect public health and the company is unwilling to conduct it in a voluntary mode.

The first steps of a recall, where FDA is involved, may include:

- A company discovers a problem and contacts FDA.
- FDA inspects a manufacturing facility, reviews records and determines the potential for a recall. State government officials can also become involved.
- FDA investigates reports of health problems through various reporting systems.
- If epidemiological data is available, the Centers for Disease Control and Prevention (CDC) contacts FDA.

The possibility of food being used as a vehicle for biological, chemical, or physical agents of harm has been recognized for several years, however; the threat of terrorism is more real today than ever, and the food supply may be a target for such an attack. With the thoughts of preparation and precaution in mind, a modern food facility must have a recall plan that will meet the expectations of federal regulatory authorities, apply to domestic and international commerce, and accomplish the task of recalling adulterated, misbranded, and/or hazardous foods. This is regardless of the cause, such as inadvertent contamination, faulty processing, hazardous food due to economic adulteration, product tampering, or even terrorism (see also the listing below).

FDA has implemented a Reportable Food Registry (RFR), which was established in 2007. The Registry is an electronic portal for industry to report when there is reasonable probability that an article of food will cause serious adverse health consequences. FDA says the Registry will help the FDA better protect public health by tracking patterns and targeting inspections.

The criterion for reporting to the RFR is:

> *Companies are required to report when there is a reasonable probability that the use of, or exposure to, an article of food will cause serious adverse health consequences or death to humans or animals.*

When decisions are being made regarding the status of a potential recall situation, the company should also consider the need to make a report using the RFR. A defective food that most likely will become a Class I recall would be a likely candidate for being reported in the RFR. Companies make their decisions based on internal and external information about the need for recalls and the need to report using the RFR. In cases of minor defects or product not yet in the marketplace, a company may implement a "Market Withdrawal" or "Stock Recovery" rather than a recall as an internal effort to recover product. Also, minor defects and/or quality problems would not warrant a report to the RFR.

Once a recall is underway, FDA will assign a Class level (see below) to the recall and expects follow-up, requiring a company to do effectiveness checks to verify the recalled product is covering the distribution chain of the product and that it is being adequately recovered.

The following is a list of possible reasons a food could be recalled. It is not all-inclusive but provides a guide to common food safety problems, and some manufacturing defects, warranting a possible recall:

Allergens – unlabeled allergenic ingredient.

Bacterial contamination – presence of harmful bacteria, which could cause illness, infection, or intoxication.

Chemical contamination – presence of unapproved chemicals. (e.g., pesticides, herbicides, fungicides, animal drugs, residual sanitizers, industrial cleaners, or solvents).

Consumer claims – consumer complaints (real or fraudulent) of injury or illness by a product.

Foreign objects – foreign objects may be present in ingredients or from processing equipment or repairs (e.g., glass, plastic, wood, or metal fragments).
Government findings – agencies or recognized organizations (e.g., CDC, FBI, DHS, State Government Agencies, etc.) may identify a potential problem with a food and can prompt regulatory agencies and food firms to take recall actions.
Human diseases – illnesses that can be transmitted through foods.
Misbranding – labels that do not declare ingredients or misstate content or nutrient composition.
Packaging defects – such as faulty seams, seals, microscopic leaks, etc.
Processing errors – companies find problems with products through internal record review and examination of processes and decide to recall as a precaution.
Suppliers' notification – ingredients or processing equipment suppliers have reason to believe there is an identified problem with their own product or equipment.
Tampering and/or tampering threats – may be the result of a disgruntled worker deliberately causing product defects, someone in the public tampering with product on the shelves, or a terrorism attempt.
Undeclared ingredients – product that contain ingredients not listed on the label (e.g., allergens).

FSMA regulations require the following, when the hazard analysis indicates there is a hazard requiring a preventive control:

A) Establish a written recall plan for the food.
B) The plan must include procedures that describe the steps to be taken, and assign responsibility for taking those steps, to perform the following actions as appropriate to the facility:
   1) Directly notify the direct consignees of the food being recalled, including how to return or dispose of the affected food;
   2) Notify the public about any hazard presented by the food when appropriate to protect public health;
   3) Conduct effectiveness checks to verify that the recall is carried out; and
   4) Appropriately dispose of recalled food – e.g., through reprocessing, reworking, diverting to a use that does not present a safety concern, or destroying the food.

Although not required, a company may wish to do a practice or mock recall to test their readiness for a real event. Recalls are classified by FDA into one of three classes, according to the level of hazard involved:

Class I: Dangerous or defective products that predictably could cause serious health problems or death. Examples include: food found to contain botulinum toxin or food with undeclared allergens.
Class II: Products that might cause a temporary health problem, or pose only a slight threat of a serious nature. Example: presence of FD&C Yellow #5 dye in candy or product containing hazardous objects like very small pieces of plastic or wood fragments.
Class III: Products that are unlikely to cause any adverse health reaction, but that violate FDA labeling or manufacturing laws. Examples include: a minor container defect or lack of English labeling in a retail food.

As for the Food Safety Plan, FDA does not require any specific forms to be used for the Recall Plan. This chapter does not provide samples of the many forms companies can use to develop their Recall Plan since these are readily available from several sources. Readers are directed to the list of Supplemental References (below) for several useful references offering suggestions on how to develop a recall plan and sample forms to use. FDA has comprehensive guidance that outlines the information and steps, including removals and corrections, necessary to conduct an effective recall:
(see http://www.fda.gov/Safety/Recalls/IndustryGuidance/ucm129259.htm).

References

Brackett, R.E., Ocasio, W., Waters, K., Barach, J. & Wan, J., 2014, *Validation and verification: A practical, industry-driven framework developed to support the requirements of the Food Safety Modernization Act (FSMA) of 2011.* Food Protection Trends. November/December. Institute for Food Safety and Health, Illinois Institute of Technology, The National Food Laboratory LLC, and Barach Enterprises LLC, USA, pp. 410–425.
Motarjemi, Y., and Lelieveld, H. (eds.), 2013, *Food safety management: A practical guide for the food industry.* Academic Press, Elsevier Inc., London.
Taylor, M.R., 2015, *Food safety in today's global food system: An FDA perspective.* Milan EXPO: World Food Safety and Security Forum. September 22. Milan, Italy.
USA Environmental Protection Agency (EPA) (2007), *Guidance for preparing standard operating procedures (SOPs),* EPA QA/G-6. EPA/600/B-07/001.

Supplemental References for Recalls

Code of Federal Regulations, Title 21, Part 7 Subpart C 7.40 to 7.59; specifically Section 7.40, the "Recall Policy."

The food recall manual, 2004, Sponsored by the US Army Grant #DAAD 13-03-03 C (-0065): FSHN 04 -10, University of Florida, Produced at IFAS Communication Services at the University of Florida, Gainesville, Florida.

FSPCA *Preventive controls for human food: Training curriculum*, 2015, Chapter 15 – "Recall Plan," The Food Safety Preventive Controls Alliance (FSPCA), viewed 13 July 2016, from http://www.iit.edu/ifsh/alliance/index.shtml.

Successfully managing product recalls and withdrawals, 4th edn., 2014, Grocery Manufacturers Association (GMA) Publication, Washington, DC.

5

How is a Food Safety System Developed and Implemented?

The success of a Food Safety System depends on preparation work done before the tasks of developing the system actually takes place. Educating and training of management and employees in the importance of their role in producing safe foods must precede the actual work. This preparatory activity should also include gathering information about the potential foodborne hazards and controls related to the products being made. It is important to recognize that employees must first understand what food safety is and then learn the skills necessary to make it function properly. Specific training and awareness activities should include management, food safety plan developers (the food safety team), and line workers responsible for aspects of food safety. Management must provide resources and adequate time for thorough education and training. Effective training and awareness of food safety hazards is an important prerequisite to the successful implementation of a Food Safety System.

As discussed previously, building a Food Safety System requires that several essential elements be in place and functioning properly. A solid foundation of programs that manage the basic environment where safe and wholesome food can be produced is required. These programs set the stage for the facility and its workers to be able to practice Good Manufacturing Practices (GMPs). Without these GMPs in place, the facility would not be able to have a safe and wholesome operating environment, and as a consequence, the high failure rate of preventive controls would make them ineffective. GMP principles and practices are implemented through a series of prerequisite programs, such as cleaning and sanitation programs. These programs are managed through the use of written standard operating procedures (SOPs). Prerequisite programs are part of the Food Safety System but do not generally appear in the Food Safety Plan. This simplifies the plan to highlight those elements that are essential to control hazards. If a plan is too complex, the importance of controlling hazards may become diluted among other activities such as basic sanitation and quality issues. A Food Safety Plan includes a comprehensive hazard

FSMA and Food Safety Systems: Understanding and Implementing the Rules,
First Edition. Jeffrey T. Barach.

analysis and appropriately designed preventive controls. Building a plan is overseen by a Preventive Controls Qualified Individual (PCQI). The plan is the playbook for the facility to make sure the system is implemented and working properly.

In summary, the Food Safety System includes management education and support, setting a food safety culture, preliminary food safety training, the establishment and maintenance of GMPs, the use of one or more PCQIs to develop the Food Safety Plan, operating the plan to ensure it is working, and making adjustments to the plan as needed. Management has a major role here to oversee the development, implementation, and operation of both the Food Safety Plan and the Food Safety System.

5.1 Developing a Food Safety Plan

As said previously, the Food Safety Plan is the playbook for the Food Safety System. The remaining part of this chapter will focus on the activities associated with plan development. There are both preliminary tasks to take before plan development, and there are specifications for developing the plan.

The format for the Food Safety Plan can vary. There is no requirement to use any specific forms or templates. In many cases the plans will describe a product and process specific and unique to the particular facility. However, some plans may use a unit operations approach. Generic plans, such as shown in a later chapter, can serve as useful guides in the development of process and product Food Safety Plans; however, it is essential that the unique conditions within each facility be considered during the development of all components of the plan. Generic plans should not be adopted as written but can serve as an outline for development of a company-specific plan. In some cases, a plan may be applicable to several similar products. FDA allows plans to include similar products and be grouped if appropriate.

Similar to a HACCP plan, in the development of a Food Safety Plan, five preliminary tasks need to be accomplished before the application of the Risk-Based Hazard Analysis and Preventive Controls can be applied to a specific product and process. The five preliminary tasks are given in Figure 1.

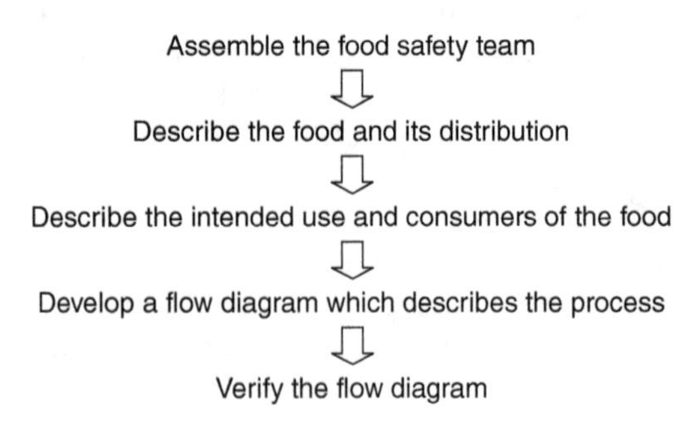
Assemble the food safety team
Describe the food and its distribution
Describe the intended use and consumers of the food
Develop a flow diagram which describes the process
Verify the flow diagram

Figure 1 Preliminary tasks in the development of the food safety plan.

5.2 Assemble the Food Safety Team

The first task in developing a Food Safety Plan is to assemble a Food Safety team consisting of individuals who have specific knowledge and expertise appropriate to the product and process. It is the team's responsibility to develop the Food Safety plan. The team should be multidisciplinary and include individuals from areas such as engineering, production, sanitation, quality assurance, and food microbiology. The team should also include local personnel who are involved in the operation, as they are more familiar with the variability and limitations of the operation. In addition, this fosters a sense of ownership among those who must implement the plan. The Food Safety team may need assistance from outside experts who are knowledgeable in the potential biological, chemical, and/or physical hazards associated with the product and the process. However, a plan that is developed totally by outside sources may be erroneous, incomplete, and lacking in support at the local level.

Due to the technical nature of the information required for hazard analysis, it is recommended that experts who are knowledgeable in the food process should either participate in or verify the completeness of the hazard analysis and the Food Safety Plan. Such individuals should have the knowledge and experience to correctly (a) conduct a hazard analysis; (b) identify potential hazards; (c) identify hazards that must be controlled; (d) recommend controls, critical limits, and procedures for monitoring and verification; (e) recommend appropriate corrective actions when a deviation occurs; (f) recommend research related to the Food Safety Plan if important information is not known; and (g) validate the Food Safety Plan.

5.3 Describe the Food and its Distribution

The Food Safety team first describes the food. This consists of a general description of the food, ingredients, and processing methods. The method of distribution should be described along with information on whether the food is to be distributed frozen, refrigerated, or at ambient temperature.

5.4 Describe the Intended Use and Consumers of the Food

Describe the normal expected use of the food. The intended consumers may be the general public or a particular segment of the population (e.g., infants, immunocompromised individuals, the elderly, etc.).

5.5 Develop a Flow Diagram that Describes the Process

The purpose of a flow diagram is to provide a clear, simple outline of the steps involved in the process. The scope of the flow diagram must cover all the steps in the process that are directly under the control of the establishment. In addition, the flow diagram can include steps in the food chain that are before and after the processing that occurs in the establishment. The flow diagram need not be as complex as engineering drawings. A block type flow diagram is sufficiently descriptive (see page 70). Also, a simple schematic of the facility is often useful in understanding and evaluating product and process flow.

5.6 Verify the Flow Diagram

The Food Safety team should perform an on-site review of the operation to verify the accuracy and completeness of the flow diagram. Modifications should be made to the flow diagram as necessary and documented.

After these five preliminary tasks have been completed, the seven principles of Food Safety are applied.

5.7 Conduct a Hazard Analysis

After addressing the preliminary tasks discussed above, the Food Safety team conducts a hazard analysis and identifies appropriate control measures. The purpose of the hazard analysis is to develop a list of hazards that are of such significance that they are reasonably likely to cause injury or illness if not effectively controlled. Hazards that are not reasonably likely to occur would not require further consideration within a Food Safety plan. It is important to consider in the hazard analysis the ingredients and raw materials, each step in the process, product storage and distribution, and final preparation and use by the consumer. When conducting a hazard analysis, safety concerns must be differentiated from quality concerns. A hazard is defined as a biological, chemical or physical agent that is reasonably likely to cause illness or injury in the absence of its control. Thus, the word *hazard* as used in this document is limited to safety.

A thorough hazard analysis is the key to preparing an effective Food Safety Plan. If the hazard analysis is not done correctly and the hazards warranting control within the Food Safety System are not identified, the plan will not be effective regardless of how well it is followed.

The hazard analysis and identification of associated control measures accomplish three objectives: Those hazards and associated control measures are identified. The analysis may identify needed modifications to a process or product so that product safety is further assured or improved. The analysis provides a basis for determining preventive controls.

In practice, the process of conducting a hazard analysis typically involves two stages. The first, hazard identification, based on risk of illness or injury, can be regarded as a brain storming session. During this stage, the food safety team reviews the ingredients used in the product, the activities conducted at each step in the process and the equipment used, the final product and its method of storage and distribution, and the intended use and consumers of the product. The biological, chemical, and physical risks posed by each ingredient or step in the process are considered. Recall that decisions made within Food Safety Modernization Act of 2011 (FSMA) regulations are risk-based for identification of potential hazards. Based on this review of risk, the team develops a list of potential biological, chemical, or physical hazards that may be introduced, increased, or controlled at each step in the production process. A knowledge of any historical adverse health-related events associated with the product and/or the process will be of value in this exercise. This step is guided by an understanding of the level of risk involved.

After the list of potential hazards is assembled, stage two, the hazard evaluation, is conducted. In stage two of the hazard analysis, the food safety team decides which potential hazards must be addressed in the Food Safety Plan. During this stage, each potential hazard is evaluated based on two criteria – the severity of the potential hazard and its likely occurrence. Severity is the seriousness of the consequences of exposure to the hazard. Considerations of severity (e.g., impact of sequelae, and magnitude and duration of illness or injury) can be helpful in understanding the public health impact of the hazard. Consideration of the likely occurrence is usually based upon a combination of experience, epidemiological data, and information in the technical literature. When conducting the hazard evaluation, it is helpful to consider the likelihood of exposure and severity of the potential consequences if the hazard is not properly controlled.

During the evaluation of each potential hazard, the food, its method of preparation, transportation, storage, and persons likely to consume the product should be considered to determine how each of these factors may influence the likely occurrence and severity of the hazard being controlled. Potential hazards that meet the criteria of being severe and likely to occur will be chosen as part of the plan and will require further evaluation. They may be considered as being managed by prerequisite programs or other means and therefore not require a preventive control. Those potential hazards not being managed will need to be mitigated by one or more preventive control measure. Those hazards that are known or reasonably foreseeable will require a preventive control. There may be differences of opinion, even among experts, as to the likely occurrence and severity of a hazard and as to the rationale provided that a potential hazard is managed by a prerequisite program or not. The food safety team may have to rely upon the opinion of experts who assist in the development of the Food Safety Plan. Also, hazards identified in one operation or facility may not be significant in another operation producing the same or a similar product. A summary of the food safety team's deliberations and the rationale developed during the hazard analysis should be written and available for future reference. This information will be useful during future reviews and updates of the hazard analysis and the Food Safety Plan.

Hazard Analysis: From Part 117.130
You must conduct a written hazard analysis to identify and evaluate, based on experience, illness data, scientific reports, and other information, known or reasonably foreseeable hazards for each type of food manufactured, processed, packed, or held at your facility to determine whether there are any hazards requiring a preventive control. Known or reasonably foreseeable hazards include: • Biological hazards, including microbiological hazards such as parasites, environmental pathogens, and other pathogens • Chemical hazards, including radiological hazards, substances such as pesticide and drug residues, natural toxins, decomposition, unapproved food or color additives, and food allergens • Physical hazards (such as stones, glass, and metal fragments) Include known or reasonably foreseeable hazards that may be present in the food for any of the following reasons: The hazard occurs naturally; the hazard may be unintentionally introduced; or the hazard may be intentionally introduced for purposes of economic gain. Note: Additional regulations may be applicable, see final rules.

5.8 Essential Elements of the Food Safety Plan

A written Food Safety Plan is required by FDA are described in FSMA in 21 CFR Part 117.126:

Food Safety Plan: From Part 117.126
• You must prepare, or have prepared, and implement a written food safety plan. • The food safety plan must be prepared, or its preparation overseen, by one or more preventive controls qualified individuals. • The written food safety plan must include: o The written hazard analysis as required by 117.130(a)(2); o The written supply-chain program as required by subpart G

- The written preventive controls as required by 117.135(b);
- The written recall plan as required by 117.139(a); and
- The written procedures for monitoring the implementation of the preventive controls as required by 117.145(a)(1);
- The written corrective action procedures as required by 117.150(a)(1); and
- The written verification procedures as required by 117.165(b).
- Records as required by this section is a record that is subject to the requirements of Subpart F of this part.

Note: Additional regulations may be applicable, see final rules.

6

What Triggers a Reanalysis of the Food Safety Plan?

A Food Safety Plan represents a point in time when the processing plan was set, the hazard analysis was done and the plan written. One must expect, however, that from that point forward there may be internal and/or external circumstances or situations that would cause one to consider the need to reevaluate or reanalyze the plan. It is very likely that a food safety system will change with time and that as a part of the system verification process, a reanalysis must be performed. FSMA rules outlines (in Part 117.170) the scope and timing of a reanalysis:

When to Conduct a Reanalysis

Conduct a reanalysis of the food safety plan as a whole at least once every three years.

Or sooner, for the whole Food Safety Plan or part of the plan:

- Whenever a significant change in the activities conducted at a facility creates a reasonable potential for a new hazard or creates a significant increased likelihood in a previously identified hazard.
- Whenever there is new information about potential hazards associated with the food.
- Whenever appropriate after an unanticipated food safety problem.
- Whenever a preventive control, combination of preventive controls, or the food safety plan as a whole is ineffective.

If a reanalysis is warranted, complete the reanalysis and then validate the new/modified preventive control (as appropriate to the nature of the preventive control). Do this:

- Before any change in activities at the facility ; or
- Within 90 calendar days after food production begins; or
- Within a reasonable timeframe with written justification (if >90 days)

FSMA and Food Safety Systems: Understanding and Implementing the Rules,
First Edition. Jeffrey T. Barach.

Reanalysis Procedures

- A PCQI (preventive controls qualified individual) must perform (or oversee) the reanalysis.
- Conduct a reanalysis of the Food Safety Plan when FDA determines it is necessary to respond to new hazards and developments in scientific understanding.
- Revise the written food safety plan if a significant change in the activities conducted at a facility creates a reasonable potential for a new hazard or a significant increase in a previously identified hazard, or document the basis for the conclusion that no revisions are needed.

Significant changes may occur in the ingredients, the process, the product, the hazard analysis conclusions, consumer claims, from scientific groups or from the marketplace that would indicate a reanalysis is appropriate.

Circumstances Suggesting a Reanalysis May Be Needed

- New information concerning the safety of the product (e.g., a regulatory alert)
- Changes in ingredients, raw materials, or their suppliers (e.g., unapproved supplier)
- Modifications in the process (e.g., equipment changes)
- Changes in the product (e.g., final moisture content)
- Finding during records review that indicates a problem (e.g., deviations and reoccurring corrective actions)
- Finding from review of consumer complaints (e.g., illnesses reported)
- New illness/injury information about similar products (e.g., scientific findings or recalls)
- New distribution procedures, new consumer handling practices, or new target population

The reanalysis could result in the need to modify or update the Food Safety Plan's preventive controls. If this involves process preventive controls, the new/modified control measure should be validated to ensure the hazard is properly mitigated. Modifications to the Food Safety Plan, preventive controls SOPs, or other changes should be approved by the Preventive Controls Qualified Individual (PCQI), and a revised plan should replace the existing plan. It is important that out-of-date plans be replaced as quickly as possible and that training on new procedures begin at once. When a reanalysis is complete, the Food Safety Team should issue a report giving details of their findings, their conclusions, and the changes they made to the Food Safety Plan.

7

Resources for Preparing Food Safety Preventive Controls Plans

7.1 Examples of Prerequisite Programs

Just as with Hazard Analysis Critical Control Points (HACCP) programs, the production of safe food products under the rules of the Food Safety Modernization Act of 2011 (FSMA) requires that the Food Safety System be built upon a solid foundation of prerequisite programs. Each segment of the food industry must provide the conditions necessary to protect food while it is under their control. This has traditionally been accomplished through the application of current good manufacturing practices (cGMPs). These conditions and practices are now considered to be prerequisite to the development and implementation of effective Food Safety Plans. Prerequisite programs, found in the FSMA rules in 21 CFR Part 117 Part B, provide the basic environmental and operating conditions that are necessary for the production of safe, wholesome food. Common prerequisite programs may include, but are not limited to:

Facilities.
The establishment should be located, constructed, and maintained according to sanitary design principles. There should be linear product flow and traffic control to minimize cross-contamination from raw to cooked materials.

Supplier Control.
Each facility should assure that its suppliers have in place effective cGMP and food safety programs. These may be the subject of continuing supplier guarantee; however, when the supplier is the one controlling the hazard (as identified in the

FSMA and Food Safety Systems: Understanding and Implementing the Rules,
First Edition. Jeffrey T. Barach.

hazard analysis), then a Supplier Preventive Control is necessary in addition to supplier's effective cGMPs. See section 7.5 below for additional information on Supplier Preventive controls.

Specifications.
There should be written specifications for all ingredients, products, and packaging materials.

Production Equipment.
All equipment should be constructed and installed according to sanitary design principles. Preventive maintenance and calibration schedules should be established and documented.

Cleaning and Sanitation.
All procedures for cleaning and sanitation of the equipment and the facility should be written and followed. A master sanitation schedule should be in place.

Personal Hygiene.
All employees and other persons who enter the manufacturing plant should follow the requirements for personal hygiene.

Training.
All employees should receive documented training in personal hygiene, Good Manufacturing Practices (GMP), cleaning and sanitation procedures, personal safety, and their role in the HACCP program.

Chemical Control.
Documented procedures must be in place to assure the segregation and proper use of nonfood chemicals in the plant. These include cleaning chemicals, fumigants, and pesticides or baits used in or around the plant.

Receiving, Storage, and Shipping.
All raw materials and products should be stored under sanitary conditions and the proper environmental conditions such as temperature and humidity to assure their safety and wholesomeness.

Traceability and Recall.
All raw materials and products should be lot-coded with a recall system in place, so that rapid and complete traces and recalls can be done when a product retrieval is necessary.

Pest Control.
Effective pest control programs should be in place.

Other examples of prerequisite programs might include quality assurance procedures; standard operating procedures for sanitation, processes, product formulations and recipes; glass control; procedures for receiving, storage, and shipping; labeling; and employee food and ingredient handling practices.

7.2 Examples of Allergen Preventive Controls

If one or more ingredients contains any of the eight major food allergens, there are two major considerations that must be addressed in the hazard analysis. The hazard analysis will allow the Preventive Controls Qualified Individual (PCQI) and the Food Safety Team to determine if an allergen hazard exists that must be controlled with an Allergen Preventive Control. First, does the possibility of cross-contact exist? This needs to be addressed at receiving, at storage, and at use of that ingredient or raw material. Next, the Team determine if a labeling hazard could exist in the operation. Labels must be checked at receipt, segregated at storage, and used so that the correct label is applied to the food. That means foods without the allergen are labeled properly and those that contain allergens have labeling to indicate their allergenic content. The two major allergen hazards to be addressed are:

Cross-contact.
Labeling.

7.3 Examples of Sanitation Preventive Controls

Sanitation Preventive Controls come into play when there is a hazard identified in the process due to cross-contamination and/or cross-contact by allergen materials. This dual role for Sanitation Preventive Controls is most evident for products that are 1) Ready-to-eat foods (since they do not receive a "kill step" is the production process or by the consumer – such as cooking) where they are susceptible to cross-contamination, and 2) Foods where the hazard analysis indicates cross-contact between products (with and without allergens) could occur – such as shared processing equipment.

Sanitation

1) Can sanitation have an impact upon the safety of the food that is being processed?
2) Can the facility and equipment be easily cleaned and sanitized to permit the safe handling of food?
3) Is it possible to provide sanitary conditions consistently and adequately to assure safe foods?

Facility design

1) Does the layout of the facility provide an adequate separation of raw materials from ready-to-eat (RTE) foods if this is important to food safety? If not, what hazards should be considered as possible contaminants of the RTE products?
2) Is positive air pressure maintained in product packaging areas? Is this essential for product safety?
3) Is the traffic pattern for people and moving equipment a significant source of contamination?

Sanitation Preventive Controls focus on:

Cross-contamination
Cross-contact

7.4 Examples of Process Preventive Controls

The hazard analysis consists of asking a series of questions that are appropriate to the process under consideration. The purpose of the questions is to assist in identifying potential hazards.

A) Ingredients
   1) Does the food contain any sensitive ingredients that may present microbiological hazards (e.g., *Salmonella, Staphylococcus aureus*); chemical hazards (e.g., aflatoxin, antibiotic or pesticide residues); or physical hazards (stones, glass, metal)?
   2) Are potable water, ice, and steam used in formulating or in handling the food?
   3) What are the sources (e.g., geographical region, specific supplier)
B) Intrinsic Factors – Physical characteristics and composition (e.g., pH, type of acidulants, fermentable carbohydrate, water activity, preservatives) of the food during and after processing.
   1) What hazards may result if the food composition is not controlled?
   2) Does the food permit survival or multiplication of pathogens and/or toxin formation in the food during processing?
   3) Will the food permit survival or multiplication of pathogens and/or toxin formation during subsequent steps in the food chain?
   4) Are there other similar products in the market place? What has been the safety record for these products? What hazards have been associated with the products?
C) Procedures used for processing
   1) Does the process include a controllable processing step that destroys pathogens? If so, which pathogens? Consider both vegetative cells and spores.

   2) If the product is subject to recontamination between processing (e.g., cooking, pasteurizing) and packaging, which biological, chemical or physical hazards are likely to occur?

D) Microbial content of the food
   1) What is the normal microbial content of the food?
   2) Does the microbial population change during the normal time the food is stored prior to consumption?
   3) Does the subsequent change in microbial population alter the safety of the food?
   4) Do the answers to the above questions indicate a high likelihood of certain biological hazards?

E) Facility design
   1) Does the layout of the facility provide an adequate separation of raw materials from ready-to-eat (RTE) foods if this is important for food safety? If not, what hazards should be considered as possible contaminants of the RTE products?
   2) Is positive air pressure maintained in product packaging areas? Is this essential for product safety?
   3) Is the traffic pattern for people and moving equipment a significant source of contamination?

F) Equipment design and use
   1) Will the equipment provide the time-temperature control that is necessary for safe food?
   2) Is the equipment properly sized for the volume of food that will be processed?
   3) Can the equipment be sufficiently controlled so that the variation in performance will be within the tolerances required to produce a safe food?
   4) Is the equipment reliable, or is it prone to frequent breakdowns?
   5) Is the equipment designed so that it can be easily cleaned and sanitized?
   6) Is there a chance for product contamination with hazardous substances; e.g., glass?
   7) What product safety devices are used to enhance consumer safety?

      metal detectors
      magnets
      sifters
      filters
      screens
      thermometers
      bone removal devices
      dud detectors

8) To what degree will normal equipment wear affect the likely occurrence of a physical hazard (e.g., metal) in the product?
9) Are allergen protocols needed in using equipment for different products?

G) Packaging
1) Does the method of packaging affect the multiplication of microbial pathogens and/or the formation of toxins?
2) Is the package clearly labeled "Keep Refrigerated" if this is required for safety?
3) Does the package include instructions for the safe handling and preparation of the food by the end user?
4) Is the packaging material resistant to damage, thereby preventing the entrance of microbial contamination?
5) Are tamper-evident packaging features used?
6) Is each package and case legibly and accurately coded?
7) Does each package contain the proper label?
8) Are potential allergens in the ingredients included in the list of ingredients on the label?

7.5 Examples of Supplier Controls

Supplier Control.
Each facility should assure that its suppliers have in place effective good manufacturing practices (GMPs) and food safety programs. These may be the subject of continuing supplier guarantee and supplier HACCP system verification.

Specifications.
There should be written specifications for all ingredients, products, and packaging materials.

7.6 Useful Forms

FSMA rules do not specify specific forms that must be used for the hazard analysis or the Food Safety Plan. Companies can use ones of their own design or ones that are readily available from government organizations or other providers, such as the Food Safety Preventive Controls Alliance (FSPCA). The following forms are ones the author has found useful – these are also used in the example Food Safety Plans in Chapter 8.

A written Food Safety Plan is required by regulation. This plan includes other written components, such as the hazard analysis, the preventive controls (including supply chain controls), monitoring procedures, corrective action plans, verification procedures, and the recall plan.

Forms in the written Food Safety Plan must have the following records information:

- The owner, operator, or agent in charge of the facility must sign and date the food safety plan.
- Upon initial completion and upon any modification.
- Provide information adequate to identify the plant or facility (e.g., the name, and when necessary, the location of the plant or facility).
- Provide the date and, when appropriate, the time of the activity documented.
- The signature or initials of the person performing the activity; and where appropriate, the identity of the product and the lot code,

Example forms are provided for:

Product Description
Flow Diagram
Hazard Analysis Worksheet (Part 1) Ingredients, Raw Materials and Packaging
Hazard Analysis Worksheet (Part 2) Processing Steps and Product Handling
Food Safety Plan Summary: Process Controls and Sanitation Controls*
Food Safety Plan Summary: Allergen Controls, Supplier controls and Other Controls*

* Some of these forms may be populated or may be blank if no specific hazards are identified. When no hazards are identified in the hazard analysis, use this example statement for that form.

Note: There were no hazards requiring preventive controls identified in the Hazard Analysis for this product; therefore, there is no Food Safety Plan Summary shown on this form.

Product Description

Product Description/ Product Name	XYZ Food Product
Product Description (including any important food safety characteristics)	
Ingredients and Raw Materials	
Packaging Used	
Intended Use	
Intended Consumers	
Shelf Life	
Labeling Instructions	
Storage and Distribution	

Flow Diagram: Outline of Typical Processing Steps

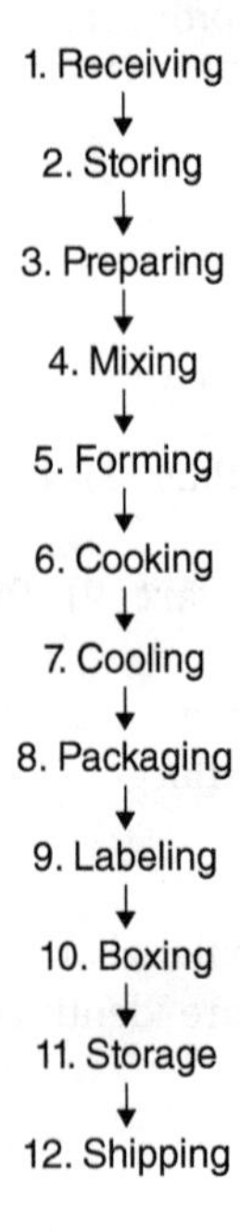

Hazard Analysis Worksheet: Ingredients, Raw Materials and Packaging (Part 1 of 2)
Product:

(1)	(2)	(3)		(4)	(5)	(6)	
Ingredient, raw material or packaging at: receiving, storage or use	Identify *potential* food safety hazards – introduced, controlled, enhanced or intentionally introduced for economic gain Biological, Chemical (including radiological) or Physical	Do any *potential* food safety hazards that are likely to occur require a preventive control be applied?		Why? Justify your decision for column 3	What preventive control(s) can be applied to prevent, eliminate, or significantly minimize the hazard? *E.G.* Process Control (including CCPs), Allergen Control, Sanitation Control, Supplier Control, or other Preventive Control	Is the preventive control applied at this step?	
		Yes*	No			Yes	No
Ingredient ABC	B? C? P?						
Raw Material BCD	B? C? P?						
Pkg. / Label CDE	B? C? P?						

* If yes, you must identify a preventive control in column 5.

Hazard Analysis Worksheet: Processing Steps and Product Handling (Part 2 of 2)
Product:

(1)	(2)	(3)		(4)	(5)	(6)	
Processing step or product handling *E.G.* heating, cooling, refrigeration, freezing, pH control, zoning, staging, rework, storage, distribution, etc.	Identify *potential* food safety hazards – introduced, controlled, enhanced, or intentionally introduced for economic gain Biological, Chemical (including radiological) or Physical	Do any *potential* food safety hazards that are likely to occur require a preventive control to be applied?		Why? Justify your decision for column 3	What preventive control(s) can be applied to prevent, eliminate or significantly minimize the hazard? *E.G.* Process Control (including CCPs), Allergen Control, Sanitation Control, Supplier Control, or other Preventive Control	Is the preventive control applied at this step?	
		Yes*	No			Yes	No
Step #1	B? C? P?						
Step #2	B? C? P?						
Handling #3	B? C? P?						

* If yes, you must identify a preventive control in column 5.

FOOD SAFETY PLAN SUMMARY: PROCESS CONTROLS AND SANITATION CONTROLS

PRODUCT:

Preventive Controls: Critical Control Point (CCP) or Sanitation Preventive Control	Hazard(s) to be Addressed in Food Safety Plan	Critical Limits for Each Control Measure	Monitoring				Corrective Action	Verification Activities	Record-keeping Procedures
			What	How	Frequency	Who			

FOOD SAFETY PLAN SUMMARY: ALLERGEN CONTROLS, SUPPLIER CONTROLS AND OTHER CONTROLS
PRODUCT:

Hazard(s) Identified in the Hazard Analysis to be Addressed and the [Preventive Control Measure Applied]	Performance Criteria (may include limits)	Monitoring (what, how & frequency)	Corrective Action	Verification Activities	Record-keeping

7.7 FSMA Training and the Food Safety Preventive Controls Alliance

The FSPCA, initiated in 2011 and coordinated by Illinois Institute of Technology's Institute for Food Safety and Health, developed a standardized training and education program and technical information network to help the domestic and foreign food industry comply with the requirements of the Preventive Controls rules for human and animal food. This work resulted in two levels of standardized hazard analysis and preventive controls training courses and distance education modules that were developed for human foods. One for students (industry and regulators) and one for trainers to become instructors.

The first "FSPCA Preventive Controls for Human Foods Training Curriculum" was developed in 2015, and courses were taught through the International Food Protection Training Institute (IFPTI), Battle Creek, MI.

The Current Good Manufacturing Practice, Hazard Analysis, and Risk-based Preventive Controls for Human Food regulation (referred to as the FSMA rules) are intended to ensure safe manufacturing/processing, packing, and holding of food products for human consumption in the United States. The regulation requires that certain activities must be completed by a preventive controls qualified individual (PCQI) who has "successfully completed training in the development and application of risk-based preventive controls." The course developed by the FSPCA is the "standardized curriculum" recognized by FDA. Successfully completing this course is one way to meet the requirements for a PCQI. An important FSMA requirement is for every processing facility to have a trained resource person or a PCQI, who has completed an FDA recognized curriculum course such as the one developed by the FSPCA. This person will oversee the implementation of the facility's food safety plan and other key tasks.

To facilitate the training outreach needed for implementation of the FSMA rules, the FSPCA also developed a train-the-trainer course for instructors. These trainers have been instructed in how to teach the FDA-recognized standardized curriculum by successfully completing the FSPCA Preventive Controls for Human Foods Lead Instructor training course.

For information on the FSPCA and training resources, see http://www.iit.edu/ifsh/alliance/.

For IFPTI courses and training opportunities, see http://www.ifpti.org/fspca-training.

8

Example Food Safety Plans

In adult learning, new skills are often acquired by studying examples, rather than studying words in a text book or listening to lectures. To that end, one of the primary purposes of this book is to use that "by example" approach in helping small to mid-sized companies quickly advance their knowledge and understanding of what they must do to meet the requirements of the rules laid out in the Food Safety Modernization Act of 2011 (FSMA). The author has provided four hypothetical foods and developed Food Safety Plans (FSPs) for each. The plans represent: a frozen product, a baked product, a refrigerated product, and a shelf-stable product. These are not real plans from an actual facility but rather are developed as examples to show what a plan looks like with regard to the relationships between ingredients, processing steps, equipment, facilities, personnel, hazards and preventive controls.

FSMA regulations do not specify forms that must be used in developing an FSP. Many companies may have their own or use those provided by consultants. Others may be found in various resources or adapted from those used in concert with their auditing systems (e.g., forms based on Global Food Safety Initiative (GFSI)-level standards for food safety). The author has developed these forms to be comprehensive and descriptive regarding the tasks at hand so that it becomes easier to populate them by either trained and untrained individuals. In breaking away from the classical approach of the Hazard Analysis Critical Control Points (HACCP) of combining the ingredients and processing steps together, these plans have them separated. This partitioning is carried through to include the FSP Summary. The rationale here is that FSMA is more complex than HACCP and with extra emphasis on individual preventive controls (allergens, supplier, etc.), in addition to process controls, it was determined by the author that a separation would be appropriate and more easily handled during the development of the FSP.

FSMA and Food Safety Systems: Understanding and Implementing the Rules,
First Edition. Jeffrey T. Barach.

Readers are cautioned that these generic plans were developed as examples for training purposes and are not intended to replace a processor's hazard analysis or the work of the Food Safety Team. In addition, the expertise of a preventive control qualified individual (PCQI) is an important part of developing a product-specific and facility-specific Food Safety Plan that is effective.

The following Food Safety Plans are offered as examples:

BBQ Sauce
Chocolate Chip Walnut Cookies
Deli Potato Salad
Mac & Cheese Frozen Meal

Barbeque Sauce – Example Food Safety Plan

Company Overview

This example company (Bubba's Best Inc.) is a very small firm that makes several shelf-stable sauce products (BBQ sauce, ketchup, steak sauce, and habenero hot sauce). BBQ sauce is made two days a week in one five-hour production shift, followed by two hours for wash down and equipment cleaning. The plant follows Good Manufacturing Practices (GMPs) regulations as described in 21 CFR Part 117, Subpart B. Prerequisite programs (PRP) for cleaning and sanitizing, as well as other PRPs, are written and performed by trained workers as standard operating procedures (SOPs). Workers record the results of their tasks when the SOP is completed. The company also has a written recall plan.

This Food Safety Plan covers production of Bubba's Best Barbeque Sauce.

Product Description/ Product Name	Bubba's Best Barbeque Sauce
Product Description (including any important food safety characteristics)	Barbeque sauce is a shelf-stable hot-fill-hold liquid product packaged in 20 ounce narrow-neck glass bottles with a screw-on plastic cap. The bottles are labeled with an adhesive label, and product is placed in a case (12 bottles) and warehoused, distributed, and retailed at room temperature. The pH of the sauce is 3.8.

Approved By: Bubba A. Smith *BAS*	Reviewed By: Betty R. Consultant *BRC*	
Product(s) Code: BBB 1001	Plant Name/Address: Bubba's Best 200 Broad St. Memphis, TN	
Issue Date/Time: March 1, 2016 8:00 AM	Reviewed/Revision Date: May 30, 2016	Page: 1

Product Description/ Product Name	Bubba's Best Barbeque Sauce
Ingredients and Raw Materials	Tomato paste, high fructose corn syrup, molasses, vinegar, smoke flavor, corn starch, garlic powder, salt and xanthan gum. Potable water is treated and tested per EPA requirements by the city.
Packaging Used	20 ounce glass bottles and lids. An adhesive pre-printed label is attached.
Intended Use	Ready-to-eat, shelf-stable product.
Intended Consumers	General public
Shelf Life	1 to 3 years stored at ambient temperature.
Labeling Instructions	Refrigerate after opening.
Storage and Distribution	Room temperature storage and retail distribution.

This generic Food Safety Plan was developed for example purposes only and is not intended to replace a processor's hazard analysis or the development of a facility's own Food Safety Plan.

Process Narrative

Receiving Ingredients: Ingredients and raw materials are purchased from domestic and international suppliers complying with recognized food safety and quality schemes. Ingredients are stored according to manufacturers' recommendations.

Receiving Packaging and Labels: 20 oz. narrow-neck bottles and screw caps are received in bulk. Adhesive labels are reviewed for conformance with product ingredients. Specifications require food grade material compatible for human food products.

Receiving Shelf-stable Ingredients:

- Tomato paste: Received in #10 cans from a national tomato product supplier.
- High fructose corn syrup, molasses, vinegar, smoke flavor, garlic powder, and corn starch: Received from U.S. approved suppliers.
- Xanthan gum: Received from an international ingredient broker.
- Salt: Received in bags from sole source broker. Specifications require food-grade salt.

Storing Ingredients, Packaging and Labels:

- **Packaging and label storage:** Glass bottles and caps are stored in the dry storage room in the packaging storage area, arranged by product code to avoid mixing of packaging. Packaging is used first-in-first-out and partially used shipping containers are closed during storage. Labels are stored by product codes. Partial rolls of label stock are used first.
- **Dry ingredient storage:** Tomato paste, high fructose corn syrup, molasses, vinegar, smoke flavor, corn starch, garlic powder, salt, and xanthan gum ingredients are stored in the dry storage room in the ingredient area, arranged by ingredient code number. All containers are sealed to avoid cross-contamination during storage.

Prepare Tomato Puree: Tomato paste is added to potable water in a steam-jacketed kettle, and the slurry is brought to about 80°F to aid in mixing. The process takes about twenty minutes.
Mix in All Other Ingredients: Dry and liquid ingredients are blended into the heated kettle to distribute them evenly. The kettle lid is closed. The sauce mixture is blended while observing the temperature until it reaches ≥ 195°F. This typically takes thirty minutes with constant mixing.
Bottles are Filled and Caped: 20 oz. bottles trays are filled with the hot sauce and caps screwed on. The bottles are inverted, to heat the neck and cap, for fifteen minutes.
Cooling and Inspection: Bottles are cooled in a cool water tunnel to bring the temperature to about 80°F. At the end of the cooling tunnel, a visual inspection is made for any packaging defects with the bottle or the cap.
Labeling: Labels are applied that match the batch record and the product inside the bottle.
Casing: Twelve bottles are placed in boxes. Labels on bottles are matched to the specific description and product number on the casepack.
Finished Product Storage: Finished product is transferred to warehouse storage (~70–80°F).
Product Shipping: Product is shipped in trucks to customers (local stores and restaurants). Products is also sold over the internet and shipped by USPS.

Approved By: Bubba A. Smith *BAS*	Reviewed By: Betty R. Consultant *BRC*	
Product(s) Code: BBB 1001	Plant Name/Address: Bubba's Best 200 Broad St. Memphis, TN	
Issue Date/Time: March 1, 2016 8:00 AM	Reviewed/Revision Date: May 30, 2016	Page: 3

Barbeque Sauce - Flow Diagram

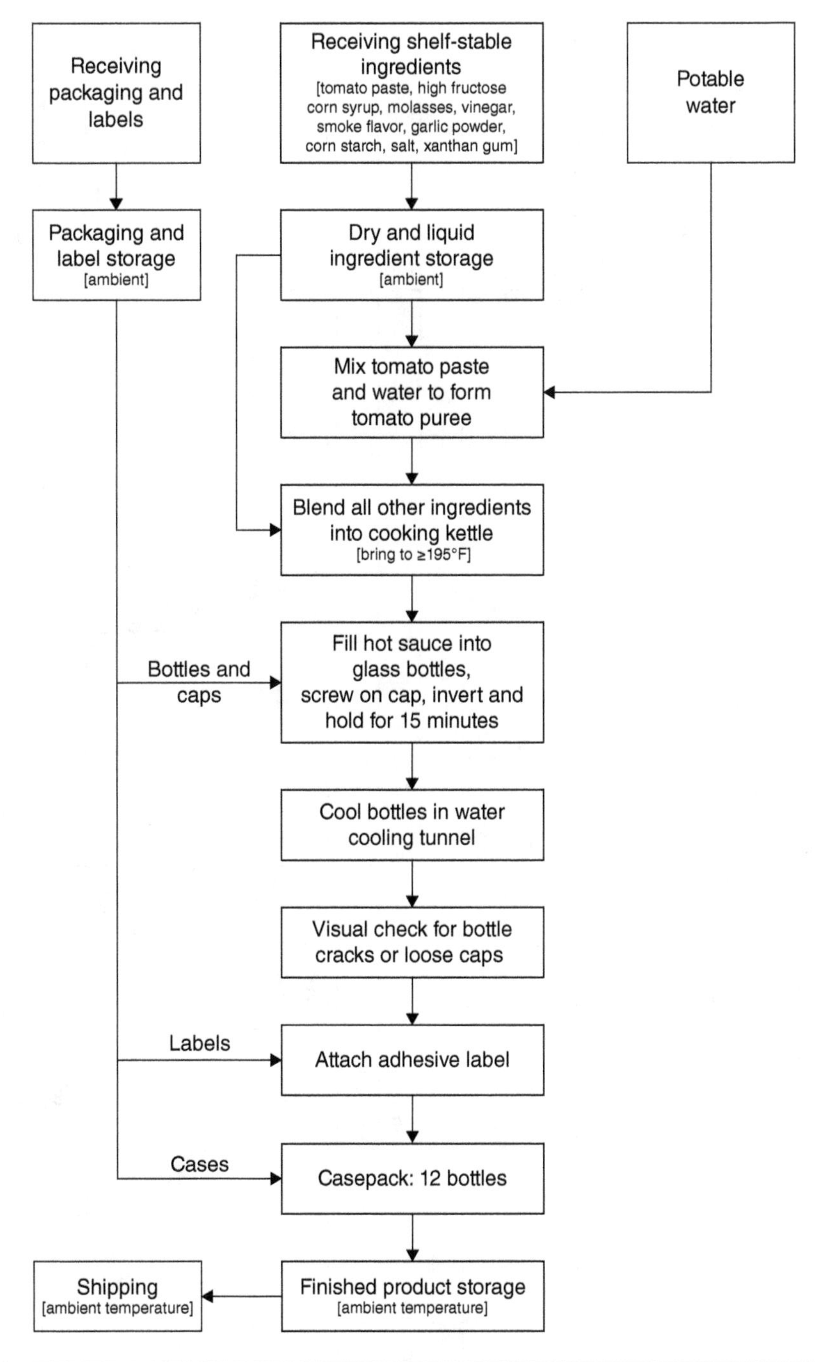

Approved By: Bubba A. Smith *BAS*	Reviewed By: Betty R. Consultant *BRC*	
Product(s) Code: BBB 1001	Plant Name/Address: Bubba's Best 200 Broad St. Memphis, TN	
Issue Date/Time: March 1, 2016 8:00 AM	Reviewed/Revision Date: May 30, 2016	Page: 4

Hazard Analysis Worksheet – Ingredients, Raw Materials and Packaging (Part 1 of 2)
Product: Barbeque Sauce

(1)	(2)		(3)		(4)	(5)	(6)	
Ingredient, raw material or packaging at: receiving, storage or use	Identify *potential* food safety hazards: introduced, controlled, enhanced or intentionally introduced for economic gain Biological, Chemical (including radiological) or Physical		Do any *potential* food safety hazards that are likely to occur require a preventive control be applied?		Why? Justify your decision for column 3	What preventive control(s) can be applied to prevent, eliminate, or significantly minimize the hazard? *E.G.* Process Control (including CCPs), Allergen Control, Sanitation Control, Supplier Control or other Preventive Control	Is the preventive control applied at this step?	
			Yes*	No			Yes	No
Receiving canned tomato paste	B?	None						
	C?	None						
	P?	None						
Receiving shelf-stable dry and liquid ingredients	B?	Vegetative pathogens such as *Salmonella* may be present in corn starch or dried spices at a low frequency. Spore-formers (e.g., *C. botulinum*) may also be in dried ingredients		X	For *Salmonella* and spore-formers the hot-fill-hold process, for this acid-type product (pH 3.8), is more than adequate to destroy any pathogens and spoilage organisms, and will prevent growth of *C. botulinum.*			
	C?	None						
	P?	None						

Approved By: Bubba A. Smith *BAS*	Reviewed By: Betty R. Consultant *BRC*
Product(s) Code: BBB 1001	Plant Name/Address: Bubba's Best 200 Broad St. Memphis, TN

Receiving labels and packaging	B?	None						
	C?	None						
	P?	Sauce bottles are glass, if broken can be a hazard		X	GMPs procedures control any potential glass breakage hazards in receiving			
Storing tomato paste	B?	None						
	C?	None						
	P?	None						
Storing other dry and liquid ingredients	B? C?	None None						
	P?	None						
Storing packaging (bottles, caps and labels)	B?	None						
	C?	None						
	P?	Sauce bottles are glass, if broken can be a hazard			GMPs procedures control any potential glass breakage hazards while in storage			

* If YES, you must identify a preventive control in column 5.

Approved By: Bubba A. Smith *BAS*	Reviewed By: Betty R. Consultant *BRC*	
Product(s) Code: BBB 1001	Plant Name/Address: Bubba's Best 200 Broad St. Memphis, TN	
Issue Date/Time: March 1, 2016 8:00 AM	Reviewed/Revision Date: May 30, 2016	Page: 6

Hazard Analysis Worksheet – Processing Steps and Product Handling (Part 2 of 2)
Product: Barbeque Sauce

(1)	(2)		(3)		(4)	(5)	(6)	
Processing step or product handling: ***E.G.* heating, cooling, refrigeration, freezing, pH control, zoning, staging, rework, storage, distribution, etc.**	**Identify *potential* food safety hazards – introduced, controlled, enhanced, or intentionally introduced for economic gain** **Biological, Chemical (including radiological), or Physical**		**Do any *potential* food safety hazards that are likely to occur require a preventive control to be applied?**		**Why? Justify your decision for column 3**	**What preventive control(s) can be applied to prevent, eliminate, or significantly minimize the hazard? *E.G.,* Process Control (including CCPs), Allergen Control, Sanitation Control, Supplier Control, or other Preventive Control**	**Is the preventive control applied at this step?**	
			Yes*	No			Yes	No
Making tomato puree	B?	None						
	C?	None						
	P?	None						
Mixing in other ingredients	B?	None						
	C?	None						
	P?	None						
Heating sauce mixture	B?	Vegetative pathogens such as *Salmonella and* Spore-formers (e.g., *C. botulinum*)		X	*Salmonella* and spore-formers may be present in corn starch or dried spices at a low frequency. The hot-fill-hold process (≥195°F initial temperature for 15 minutes), for this acid-type product (pH 3.8), is more than adequate to destroy any pathogens and spoilage organisms.			
	C?	None						
	P?	None						

Approved By: Bubba A. Smith *BAS*	Reviewed By: Betty R. Consultant *BRC*
Product(s) Code: BBB 1001	Plant Name/Address: Bubba's Best 200 Broad St. Memphis, TN

Fill bottles and cap, invert and hold 15 minutes	B?	None						
	C?	None						
	P?	Glass breakage/ fragments	X		Breakage may occur during capping procedures.	***Process Control*** – Glass bottle inspection and glass breakage control program at capping. Monitor capping area and implement glass breakage procedures if they occur.	X	
Cool bottles and inspect	B?	None						
	C?	None						
	P?	Glass breakage/ fragments	X		The shock of temperature change can cause glass breakage during heating and cooling.	***Process Control*** – Glass bottle inspection and glass breakage control program. Monitor cooling area and implement glass breakage procedures if they occur.	X	
Put on adhesive label	B?	None						
	C?	None						
	P?	None						

(continued)

(1)	(2)		(3)		(4)	(5)	(6)	
Processing step or product handling: *E.G.* heating, cooling, refrigeration, freezing, pH control, zoning, staging, rework, storage, distribution, etc.	Identify *potential* food safety hazards – introduced, controlled, enhanced, or intentionally introduced for economic gain Biological, Chemical (including radiological), or Physical		Do any *potential* food safety hazards that are likely to occur require a preventive control to be applied?		Why? Justify your decision for column 3	What preventive control(s) can be applied to prevent, eliminate, or significantly minimize the hazard? *E.G.,* Process Control (including CCPs), Allergen Control, Sanitation Control, Supplier Control, or other Preventive Control	Is the preventive control applied at this step?	
Case and Store finished product	B?	None						
	C?	None						
	P?	None						
Ship finished product	B?	None						
	C?	None						
	P?	None						

* If YES, you must identify a preventive control in column 5.

Approved By: Bubba A. Smith *BAS*	Reviewed By: Betty R. Consultant *BRC*
Product(s) Code: BBB 1001	Plant Name/Address: Bubba's Best 200 Broad St. Memphis, TN

FOOD SAFETY PLAN SUMMARY: PROCESS CONTROLS AND SANITATION CONTROLS

PRODUCT: Barbeque Sauce

Preventive Controls: Critical Control Point (CCP) or Sanitation Preventive Control	Hazard(s) to Be Addressed in Food Safety Plan	Critical Limits for Each Control Measure	Monitoring				Corrective Action	Verification Activities	Record-keeping Procedures
			What	How	Frequency	Who			
Glass Control Program	Glass fragments	No broken glass in capping area or in cooling inspection area	Presence of broken glass	Visual check for glass pieces	Before start of batch run, at the middle of the run and at the end of the run	Capping operator	Stop production and adjust equipment if necessary	QA supervisor reviews visual monitoring, corrective action and verification records within one week of production	Capping visual check log,
[Process preventive control]						Filler operator	Remove broken glass plus 10 bottles before and after breakage point	QA supervisor reviews consumer complaints for glass in product, monthly	Cooling visual check log.
							Segregate product since last good check and determine status and disposition of product		Corrective action records
									Consumer complaints review summary
									Training records for Q.I.s

Approved By: Bubba A. Smith *BAS*	Reviewed By: Betty R. Consultant *BRC*	
Product(s) Code: BBB 1001	Plant Name/Address: Bubba's Best 200 Broad St. Memphis, TN	
Issue Date/Time: March 1, 2016 8:00 AM	Reviewed/Revision Date: May 30, 2016	Page: 10

Chocolate Chip Walnut Cookies – Example Food Safety Plan

Company Overview

This example company (Kathy's Kookies LLC.) is a small firm that makes a variety of cookies for internet and retail sales. Products include Chocolate Chip Walnut Cookies, Peanut Butter Cookies, Plain Chocolate Chip Cookies, and Sugar Cookies. Products are made three-days a week in one 8-hour production shift, followed by two hours for clean-up and sanitation of tools and equipment. For production staging of allergens, Plain Chocolate Chip Cookies are followed, on the same day, by Chocolate Chip Walnut Cookies. On other days, only one type of cookie is made.

This Food Safety Plan covers production of Chocolate Chip Walnut Cookies. The baking cooking equipment and packaging "clean" area are used to make other products on separate days. The plant follows Good Manufacturing Processes (GMP) regulations as described in 21 CFR Part 117, Subpart B. Prerequisite programs (PRP) for cleaning and sanitizing, as well as other PRPs, are written and performed by trained workers as standard operating procedures (SOPs). Workers record the results of their tasks when the SOP is completed. The company also has a written recall plan.

Product Description/ Product Name	Chocolate Chip Walnut Cookies
Product Description (including any important food safety characteristics)	Chocolate Chip Walnut Cookies are ready-to-eat desserts packed in a 20 ounce plastic tray containers, sealed with a pealable adhesive labeled film. The product is placed 16 trays to a box and stored, warehoused, distributed, and retailed at ambient temperatures.
Ingredients and Raw Materials	Flour, shortening, sugar, eggs, baking soda, salt, vanilla, chocolate chips, chopped walnuts and soy lecithin.
Packaging Used	20 ounce plastic trays with pre-labeled adhesive film is pressed on plastic lid. Case packs are cardboard boxes.
Intended Use	For internet sales and retail stores.
Intended Consumers	General public.
Shelf Life	90 days at room temperature.
Labeling Instructions	Allergen labeling. Manufacturing date.
Storage and Distribution	Ambient storage, internet sales and retail distribution.

Approved By: Kathy A. Cook *KAW*	Reviewed By: Chip R. Goody *CRG*	
Product(s) Code: CCW 23W	Plant Name/Address: Kathy's Kookies 55 South St. Miami, FL	
Issue Date/Time: Jan. 18, 2016 9:00 AM	Reviewed/ Revision Date: Feb. 18, 2016	Page: 1

This generic Food Safety Plan was developed for example purposes only and is not intended to replace a processor's hazard analysis or the development of a facility's own Food Safety Plan.

Process Narrative

Receiving Ingredients: Ingredients and raw materials are purchased from domestic suppliers complying with recognized food safety and quality schemes. Ingredients are stored according to manufacturers' recommendations.
Receiving Packaging and Labels: 20 ounce pre-formed plastic containers, pre-printed adhesive sealable film with labeling information are received in bulk and stored in a warehouse. Pre-printed film labels are reviewed for conformance with product allergen requirements and ingredients. Specifications require food-grade material compatible for packaged products.

Receiving Shelf-Stable Ingredients:

- All-purpose flour (wheat): Received in 100 pound bags from a national baking goods supplier.
- Chopped walnuts: Purchased from a California supplier in bulk, produced by them using Good Agricultural Practices (GAP) procedures.
- Semi-sweet chocolate chips, vegetable shortening, sugar, salt, vanilla extract, baking soda : Purchased as restaurant-size containers from a local big-box store.
- Labels: Purchased pre-printed from a label supply company.
- Packaging: Trays and case boxes purchased from a national package supplier.
- Soy lecithin pan release agent: Purchased from a national ingredient supplier.

Receiving Refrigerated Ingredients:

- Eggs: Purchased refrigerated in bulk from a local big-box store.

Storing Ingredients and Packaging:

- **Packaging and Label storage:** Trays, adhesive lids with labels, and case packs are stored in the dry storage room in the packaging area, arranged by product code to avoid mixing of packaging. Allergen labels are color-coded to avoid a mix-up. Packaging is used first-in-first-out, and partially used shipping containers are closed during storage.
- **Dry ingredient and liquid storage:** Flour, shortening, sugar, baking soda, salt, vanilla extract, chocolate chips, chopped walnuts, and soy lecithin are stored in the dry storage room in the ingredient area, arranged by ingredient code number. All containers are sealed to avoid food allergen cross-contact

and cross-contamination during storage. Ingredients containing food allergens are identified and stored in specific locations with any similar allergenic ingredients. There are separated/color-coded areas for wheat flour, walnuts, and peanut butter (not in this product).
- **Refrigerated ingredient storage:** Shell eggs are stored in a refrigerator that is kept at (≤41°F) and used on a first-in-first-out basis.

Dough Preparation: Dough is prepared in two steps. In a dough mixer, shortening, sugar, eggs, and vanilla are beat for about 15 minutes. Flour and additional ingredients are blended in for the formation of cookie dough – about 20 minutes of slow mixing.

Staging Ingredients Mixing: On days when Chocolate Chip cookies are made, Plain Chocolate Chip Cookies are made first, followed by Chocolate Chip Walnut Cookies. This prevents cross-contact problems of getting walnuts in the plain cookies mix.

Portioning Dough on Baking Pans: Using a mechanical scoop, 1/4 cup of dough is portioned evenly on baking pans. Pans are sprayed with soy lecithin to prevent sticking. Any leftover dough is used as rework but is clearly labeled and stored in sealed containers.

Bake Cookies: Pans are loaded into a batch oven, and cookies are baked for 12 minutes at ≥350°F. Product temperature inside cookie (internal measurement) exceeds 180°F for one minute in this processing and cooling time.

Cool Cookies: Hot cookies on baking pans are loaded into cooling racks to reach a product temperature to ≤80°F in 30 minutes. This is in a "clean" packaging area, and workers are gowned and use one-use disposable gloves to move cookies and pack trays.

Fill, Weigh, Lid, and Labeling: 20 oz. trays are filled with cookies, weighed, and lidded with the label. Labels on the trays are matched to the specific description of the product. Manufacturing dates are provided on the label by ink-jet.

Cartoning and Casing: The trays are placed in pre-labeled cardboard cases – 16 trays to a box. Pre-printed case information describes the product.

Finished Product Storage: Finished product is transferred to ambient storage in a warehouse area.

Finished Product Shipping: Product is shipped in trucks to customers (internet customers and grocery stores).

Approved By: Kathy A. Cook *KAW*	Reviewed By: Chip R. Goody *CRG*	
Product(s) Code: CCW 23W	Plant Name/Address: Kathy's Kookies 55 South St. Miami, FL	
Issue Date/Time: Jan. 18, 2016 9:00 AM	Reviewed/ Revision Date: Feb. 18, 2016	Page: 3

CHOCOLATE CHIP WALNUT COOKIES – Flow Diagram

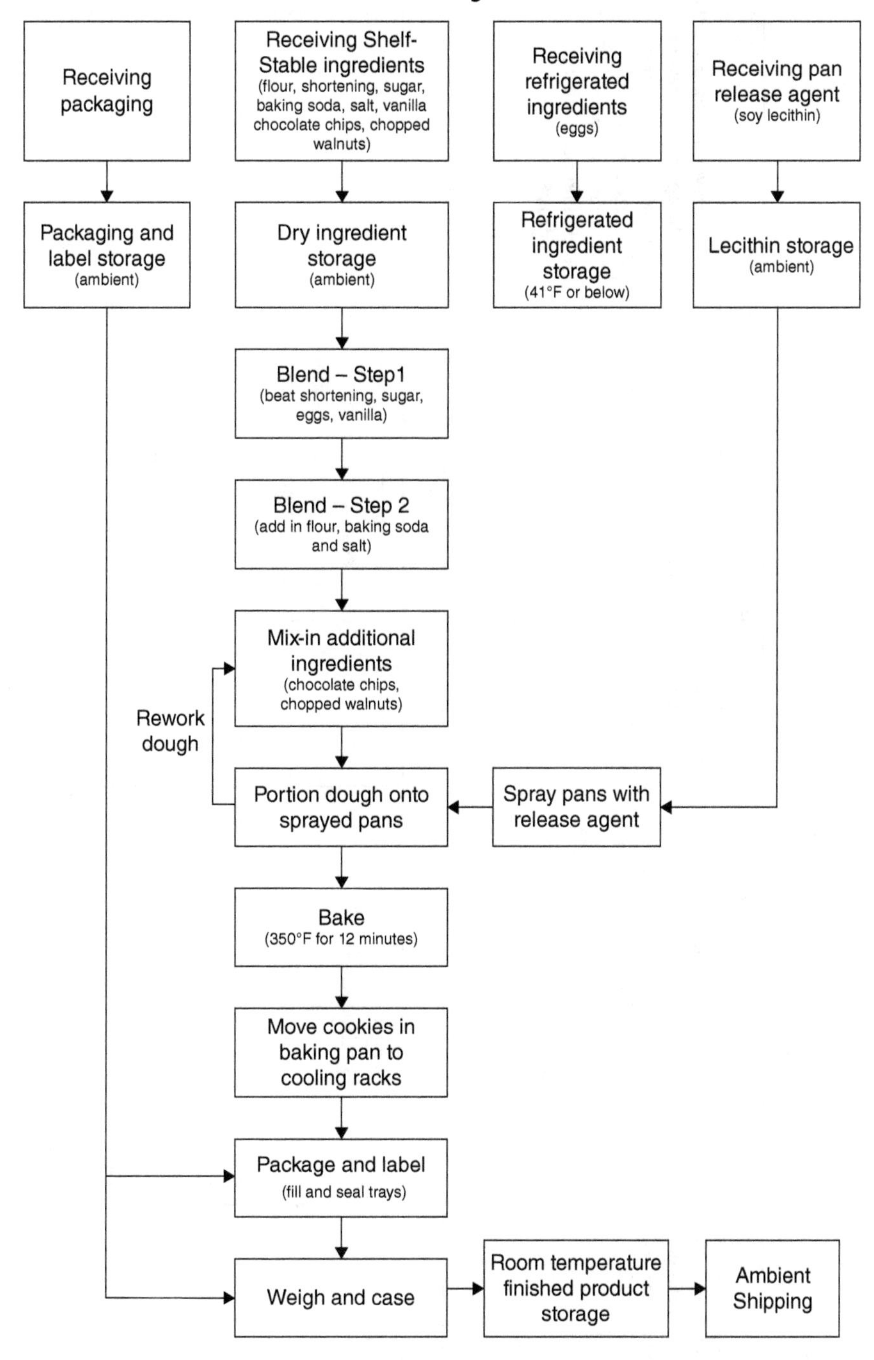

Approved By: Kathy A. Cook *KAW*	Reviewed By: Chip R. Goody *CRG*	
Product(s) Code: CCW 23W	Plant Name/Address: Kathy's Kookies 55 South St. Miami, FL	
Issue Date/Time: Jan. 18, 2016 9:00 AM	Reviewed/ Revision Date: Feb. 18, 2016	Page: 4

Hazard Analysis Worksheet – Ingredients, Raw Materials, and Packaging (Part 1 of 2)
Product: CHOCOLATE CHIP WALNUT COOKIES

(1)	(2)		(3)		(4)	(5)	(6)	
Ingredient, raw material or packaging at: receiving, storage or use	**Identify *potential* food safety hazards – introduced, controlled, enhanced, or intentionally introduced for economic gain Biological, Chemical (including radiological), or Physical**		**Do any *potential* food safety hazards that are likely to occur require a preventive control to be applied?**		**Why? Justify your decision for column 3**	**What preventive control(s) can be applied to prevent, eliminate, or significantly minimize the hazard? *E.G.,* Process Control (including CCPs), Allergen Control, Sanitation Control, Supplier Control or other Preventive Control**	**Is the preventive control applied at this step?**	
			Yes*	No			Yes	No
Receiving refrigerated shell eggs	B?	Vegetative pathogens such as *Salmonella*		X	*Salmonella,* may be present on or in eggs at a very low frequency, and subsequent baking is more than adequate to destroy them.			
	C?	Undeclared egg allergens	X		Labels must declare allergens when present, and printing errors can happen	***Allergen Control*** for allergen labeling – label review upon receipt at a later step		X
	P?	None						

Approved By: Kathy A. Cook *KAW*	Reviewed By: Chip R. Goody *CRG*	
Product(s) Code: CCW 23W	Plant Name/Address: Kathy's Kookies 55 South St. Miami, FL	
Issue Date/Time: Jan. 18, 2016 9:00 AM	Reviewed/ Revision Date: Feb. 18, 2016	Page: 5

Receiving soy lecithin	B?							
	C?	Undeclared soy allergens	X		Labels must declare allergens when present, and printing errors can happen	***Allergen Control*** for allergen labeling – label review upon receipt at a later step		X
	P?	None						
Receiving flour	B?	Vegetative bacteria such as *Salmonella* and patho-genic *E. coli* have been shown to be present in flour		X	*Salmonella* and pathogenic *E. coli* may be present in flour; however, baking to an internal temp of 180°F for one minute is more than adequate to destroy them.			
	C?	Mycotoxin deoxynivale-nol (DON)		X	Approved growers and suppliers are required to test for mycotox-ins, like DON, in order to prevent them from entering the food chain.			
	P?							

(*continued*)

(1)	(2)		(3)		(4)	(5)	(6)	
Ingredient, raw material or packaging at: receiving, storage or use	**Identify *potential* food safety hazards – introduced, controlled, enhanced, or intentionally introduced for economic gain <u>B</u>iological, <u>C</u>hemical (including radiological), or <u>P</u>hysical**		**Do any *potential* food safety hazards that are likely to occur require a preventive control to be applied?**		**Why? Justify your decision for column 3**	**What preventive control(s) can be applied to prevent, eliminate, or significantly minimize the hazard? *E.G.,* Process Control (including CCPs), Allergen Control, Sanitation Control, Supplier Control or other Preventive Control**	**Is the preventive control applied at this step?**	
Receiving dry and liquid ingredients (shortening, sugar, baking soda, salt, vanilla extract, chocolate chips, chopped walnuts)	B?	Vegetative bacteria such as *Salmonella* and patho-genic *E. coli* could be present in some dry ingredients		X	*Salmonella* and pathogenic *E. coli* may be present in other ingredients; however, baking to an internal temp of 180°F for one minute is more than adequate to destroy them.			
	C?	Undeclared nut allergens	X		Labels must declare allergens when present, and printing errors can happen	***Allergen Control*** for allergen labeling – label review upon receipt at a later step		X

Approved By: Kathy A. Cook *KAW*	Reviewed By: Chip R. Goody *CRG*
Product(s) Code: CCW 23W	Plant Name/Address: Kathy's Kookies 55 South St. Miami, FL

	P?	Foreign material in chopped walnuts (shells) may be associated with the processing of nuts.		X	Shell fragments are managed by approved suppliers, and are not likely to be present.			
Receiving packaging (trays, lids with labels)	B?	None						
	C?	Undeclared allergens (wheat, walnut, soy, egg)	X		Labels must declare allergens when present, and printing errors can happen	***Allergen Control*** for allergen labeling – label review upon receipt	X	
	P?	None						
Storing refrigerated eggs	B?	None						
	C?	None						
	P?	None						

(continued)

(1)	(2)		(3)		(4)	(5)	(6)	
Ingredient, raw material or packaging at: receiving, storage or use	Identify *potential* food safety hazards – introduced, controlled, enhanced, or intentionally introduced for economic gain Biological, Chemical (including radiological), or Physical		Do any *potential* food safety hazards that are likely to occur require a preventive control to be applied?		Why? Justify your decision for column 3	What preventive control(s) can be applied to prevent, eliminate, or significantly minimize the hazard? *E.G.*, Process Control (including CCPs), Allergen Control, Sanitation Control, Supplier Control or other Preventive Control	Is the preventive control applied at this step?	
Storing dry ingredients and shelf-stable liquid	B?	None						
	C?	Undeclared allergens (wheat, walnut, soy)	X		Ingredients where allergens are declared must be separated and stored from other ingredients where no allergens could allow cross-contact; allergen-containing ingredients are in a color coded area.	***Allergen Control*** for allergen ingredient storage – part of Allergen Control Program	X	
	P?	None						
Storing packaging materials and labels	B?	None						
	C?	Undeclared allergens (wheat, walnut, soy, egg)	X		Labels where allergens are declared must be separated and stored from other labels for products when no allergens are present; allergen labels are in a color coded area.	***Allergen Control*** for allergen label storage – part of Allergen Control Program	X	
	P?	None						

* If YES, you must identify a preventive control in column 5.

Approved By: Kathy A. Cook *KAW*	Reviewed By: Chip R. Goody *CRG*
Product(s) Code: CCW 23W	Plant Name/Address: Kathy's Kookies 55 South St. Miami, FL

Hazard Analysis Worksheet – Processing Steps and Product Handling (Part 2 of 2)
Product: CHOCOLATE CHIP WALNUT COOKIES

(1)	(2)		(3)		(4)	(5)	(6)	
Processing step or product handling. ***E.G.,* heating, cooling, refrigeration, freezing, pH control, zoning, staging, rework, storage, distribution, etc.**	**Identify *potential* food safety hazards – introduced, controlled, enhanced, or intentionally introduced for economic gain** **Biological, Chemical (including radiological) or Physical**		**Do any *potential* food safety hazards that are likely to occur require a preventive control to be applied?**		**Why? Justify your decision for column 3**	**What preventive control(s) can be applied to prevent, eliminate or significantly minimize the hazard?** ***E.G.,* Process Control (including CCPs), Allergen Control, Sanitation Control, Supplier Control, or other Preventive Control**	**Is the preventive control applied at this step?**	
			Yes*	No			Yes	No
Blend major ingredients	B?	None			.			
	C?	None						
	P?	None						
Mix in chips and nuts	B?	None						
	C?	Cookies made without nuts are made first, and second batch that day has walnuts added		X	Staging of batch done to avoid having nuts cross-contact with non-nut product is accomplished by scheduling, using SOPs and color coding.			
	P?	None						

(*continued*)

Approved By: Kathy A. Cook *KAW*	Reviewed By: Chip R. Goody *CRG*	
Product(s) Code: CCW 23W	Plant Name/Address: Kathy's Kookies 55 South St. Miami, FL	
Issue Date/Time: Jan. 18, 2016 9:00 AM	Reviewed/ Revision Date: Feb. 18, 2016	Page: 10

(1)	(2)		(3)		(4)	(5)	(6)	
Processing step or product handling. ***E.G.,*** **heating, cooling, refrigeration, freezing, pH control, zoning, staging, rework, storage, distribution, etc.**	**Identify *potential* food safety hazards – introduced, controlled, enhanced, or intentionally introduced for economic gain** **<u>B</u>iological, <u>C</u>hemical (including radiological) or <u>P</u>hysical**		**Do any *potential* food safety hazards that are likely to occur require a preventive control to be applied?**		**Why? Justify your decision for column 3**	**What preventive control(s) can be applied to prevent, eliminate or significantly minimize the hazard?** ***E.G.,*** **Process Control (including CCPs), Allergen Control, Sanitation Control, Supplier Control, or other Preventive Control**	**Is the preventive control applied at this step?**	
Portion dough on pans spayed with non-stick agent	B?	None			.			
	C?	None						
	P?	None						
Rework dough	B?	None						
	C?	Undeclared allergens	X		Rework dough where allergens are present must be separated and stored from other ingredients; allergen-containing rework is color coded to prevent cross-contact.	***Allergen Control*** for rework dough – part of Allergen Control Program	X	
	P?	None						

Approved By: Kathy A. Cook *KAW*	Reviewed By: Chip R. Goody *CRG*
Product(s) Code: CCW 23W	Plant Name/Address: Kathy's Kookies 55 South St. Miami, FL

Bake cookies	B?	Vegetative pathogens such as pathogenic *E. coli* and *Salmonella*		X	Validation data demonstrated that for cookies, internal cookie temperatures achieved during the baking process (>180°F for 1 minute) to achieve a palatable texture far exceeds those needed to destroy vegetative pathogens.			
	C?	None						
	P?	None						
Cool cookies	B?	None						
	C?	None						
	P?							
Package: fill trays, weigh, apply lid with label	B?	None						
	C?	Undeclared Allergens on label – wheat, walnuts, soy, eggs	X		Product contains allergens Product ingredients must match label.	***Allergen Control*** – declaration of allergens on label	X	
	P?	None						

(*continued*)

(1)	(2)		(3)		(4)	(5)	(6)	
Processing step or product handling. ***E.G.*, heating, cooling, refrigeration, freezing, pH control, zoning, staging, rework, storage, distribution, etc.**	**Identify *potential* food safety hazards – introduced, controlled, enhanced, or intentionally introduced for economic gain** **Biological, Chemical (including radiological) or Physical**		**Do any *potential* food safety hazards that are likely to occur require a preventive control to be applied?**		**Why? Justify your decision for column 3**	**What preventive control(s) can be applied to prevent, eliminate or significantly minimize the hazard?** ***E.G.*, Process Control (including CCPs), Allergen Control, Sanitation Control, Supplier Control, or other Preventive Control**	**Is the preventive control applied at this step?**	
Storage of finished product	B?	None						
	C?	None						
	P?	None						
Shipping of finished product	B?	None						
	C?	None						
	P?	None						

* If YES, you must identify a preventive control in column 5.

Approved By: Kathy A. Cook *KAW*	Reviewed By: Chip R. Goody *CRG*
Product(s) Code: CCW 23W	Plant Name/Address: Kathy's Kookies 55 South St. Miami, FL

FOOD SAFETY PLAN SUMMARY: ALLERGEN CONTROLS, SUPPLIER CONTROLS AND OTHER CONTROLS
PRODUCT: CHOCOLATE CHIP WALNUT COOKIES

Hazard(s) Identified in the Hazard Analysis to be Addressed and the (Preventive Control Measure Applied)	Performance Criteria (may include limits)	Monitoring (what, how & frequency)	Corrective Action	Verification Activities	Record-keeping
Wheat, Walnut, Egg and Soy Allergens *(Allergen Preventive Control Program at Receiving, storing, rework and at labeling)*	Correct labels received and stored separately from non-allergen labels Allergens stored in proper location in sealed containers and color coded Correct labels applied to finished product	Visual inspection of labels upon receipt Proper storage upon receipt of labels Proper storage of allergens containers whenever opened and used Visual inspection by line operator of labels in packaging machine before and at end of run to ensure proper allergen labeling information on label	If wrong labels are received, reject shipment. Notify supplier If labels or holding containers for rework are out of place, correct and notify supervisor If label is incorrect, the product is held in quarantine until a decision is made regarding its relabeling or its disposal	QA manager does independent check to determine labeling is correct. One example labeled product is taken for verification of proper label. QA manager checks storage and rework area. QA manager reviews and initials records daily and compares results with past reviews to determine any trends.	Monitoring records Verification records Corrective action log Employee allergen training records for Q.I.s

Approved By: Kathy A. Cook *KAW*	Reviewed By: Chip R. Goody *CRG*	
Product(s) Code: CCW 23W	Plant Name/Address: Kathy's Kookies 55 South St. Miami, FL	
Issue Date/Time: Jan. 18, 2016 9:00 AM	Reviewed/ Revision Date: Feb. 18, 2016	Page: 14

Deli Potato Salad – Example Food Safety Plan

Company Overview

This example company (Dave's Deli Delights Inc.) is a small firm that makes a variety of refrigerated deli-type products. Products include Potato Salad, Cole Slaw, Pasta Salad, and Cheese and Broccoli Pasta Salad. Products are made five days a week in one 8-hour production shift, followed by four hours for sanitation. The plant follows Good Manufacturing Procedures (GMPs) regulations as described in 21 CFR Part 117, Subpart B. Prerequisite programs (PRP) for cleaning and sanitizing, as well as other PRPs, are written and performed by trained workers as standard operating procedures (SOPs). Workers record the results of their tasks when the SOP is completed.

This Food Safety Plan covers production of Deli Potato Salad.The cooking equipment and hygienic zone area are used to make other products on separate days. The company also has a written recall plan.

Product Description/ Product Name	Deli Potato Salad
Product Description (including any important food safety characteristics)	Deli Potato Salad is a refrigerated, ready-to-eat (RTE) side dish packed in 16 ounce plastic containers (cups), sealed with an adhesive transparent film and over-capped with a plastic lid. The refrigerated product is placed eight containers to a box and warehoused, distributed, and retailed refrigerated.
Ingredients and Raw Materials	Cooked potatoes, onions, celery, parsley, sugar, mayonnaise, mustard, salt, pepper, and garlic powder. Potable water is treated and tested per EPA requirements by the city.
Packaging Used	16 ounce plastic cups with clear film under the pressed-on plastic lid. Labels are pre-printed adhesive labels.
Intended Use	For delis and retail stores that offer a ready-to-eat, refrigerated product.
Intended Consumers	General public
Shelf Life	10 days under refrigeration.
Labeling Instructions	Keep refrigerated. Allergen labeling. Use-by date.
Storage and Distribution	Refrigerated storage and retail distribution

Approved By: I.M. Dave *IMD*	Reviewed By: Josh Handy *JH*	
Product(s) Code: DPS 505	Plant Name/Address: Dave's Deli Delights 5 Broad St., Long, TX	
Issue Date/Time: Jan. 15, 2016 9:00 AM	Reviewed/Revision Date: Feb. 15, 2016	Page: 1

This generic Food Safety Plan was developed for example purposes only and is not intended to replace a processor's hazard analysis or the development of a facility's own Food Safety Plan.

Process Narrative

Receiving Ingredients: Ingredients and raw materials are purchased from domestic suppliers complying with recognized food safety and quality schemes. Ingredients are stored according to manufacturers' recommendations.

Receiving Packaging and Labels: 16 ounce containers, pre-printed adhesive labels, adhesive sealable film and press-on plastic lids are received in bulk and stored in a warehouse. Pre-printed labels are reviewed for conformance with product allergen requirements and ingredients. Specifications require food-grade material compatible for packaged refrigerated products.

Receiving Shelf Stable Ingredients:

- Potatoes, onions: Received from sole source local farm supplier.
- Pepper and garlic powder: Purchased from a national supplier in bulk.
- Mayonnaise and mustard: Purchased as restaurant-size containers from a local big-box store. Opened mayonnaise is refrigerated to help with cooling product.
- Packaging: Purchased from a national package supplier.
- Salt: Received in bags from sole source broker. Specifications require food grade salt.

Receiving Refrigerated Ingredients:

- Celery and parsley are purchased through an international broker that supplies fresh produce year-round. The country of origin is Mexico.

Storing Ingredients and Packaging:

- **Packaging storage:** Cups, film, lids, and labels are stored in the dry storage room in the packaging area, arranged by product code to avoid mixing of packaging. Packaging is used first-in-first-out and partially used shipping containers are closed during storage.
- **Dry ingredient storage:** Potatoes, onions, spices, and unopened mustard and mayonnaise are stored in the dry storage room in the ingredient area, arranged by ingredient code number. All containers are sealed to avoid food allergen cross-contact and cross-contamination during storage. Ingredients containing food allergens (mayonnaise) are identified and stored in specific locations with any similar allergenic ingredients.
- **Refrigerated ingredient storage:** Celery and parsley are stored in a cooler that is kept at ≤41°F and used on a first-in-first-out basis. Opened mayonnaise is stored refrigerated.

Approved By: I.M. Dave *IMD*	Reviewed By: Josh Handy *JH*	
Product(s) Code: DPS 505	Plant Name/Address: Dave's Deli Delights 5 Broad St., Long, TX	
Issue Date/Time: Jan. 15, 2016 9:00 AM	Reviewed/Revision Date: Feb. 15, 2016	Page: 2

Vegetable Preparation: Potatoes are washed, peeled, and cut in ~1 inch sections by hand. A hygienic zone is used to protect cut vegetables from contamination. Workers are gowned and use one-use disposable gloves. Onions are peeled, celery, and parsley are washed and trimmed, and all are chopped by hand; then rinsed in fresh potable water and shaken to remove water.

Cook Potatoes: Salt is added to potable water in a steam-jacketed kettle and is brought to boiling. The potato sections are added to boiling water and cooked for at least 15 minutes to achieve acceptable texture. Product temperature (internal measurement) exceeds 145°F in this processing time. The cooking kettle is drained, cooked potatoes are retained, and water is discarded.

Cool, Drain Potatoes: Potable cold water is used to cool the drained potatoes in ≤4 hours to ≤41°F. Municipal cold water temperatures are 50–55°F depending on the season. Potatoes are cooled by running cold water and/or adding potable ice. Generally, the potatoes are brought to a cold water temperature(≤41°F)in ~2+ hours, after which they are drained and transferred to the hygienic zoned assembly area, where it is weighed and mixed with other ingredients. A potato salad batch is typically assembled with other ingredients and packaged in <30 minutes after delivery to the hygienic assembly area.

Mix all Ingredients: Dry and liquid ingredients are gently blended in kettle mixer with potato sections and chopped vegetables. They are blended to a uniform mixture. A batch is blended in <15 minutes. The blended ingredient temperature is typically below ambient after mixing.

Fill, Weigh, Lidding, and Labeling: 16 oz. cups are filled, weighed, lidded, and labeled. Labels on the cups are matched to the specific product description, batch information, and product number. Keep Refrigerated warnings and Use-by Dates are provided on the label. Rework is always discouraged due to the fact this product has no further processing.

Refrigeration: Packaged product passes into an air-circulating refrigeration unit to quickly cool the product temperature to ≤41°F in 60 minutes or less.

Cartoning and Casing: The one pound filled cups are placed in pre-labeled cardboard cases – 8 cups per box. Case-handling information says what the product is and to keep it refrigerated.

Refrigerated Finished Product Storage: Finished product is transferred to refrigerated storage (≤41°F).

Refrigerated Product Shipping: Product is shipped in refrigerated trucks to customers (delis, convenience stores, office cafeterias, quick serve restaurants, and grocery stores) under refrigerated conditions (≤41°F).

Approved By: I.M. Dave *IMD*	Reviewed By: Josh Handy *JH*	
Product(s) Code: DPS 505	Plant Name/Address: Dave's Deli Delights 5 Broad St., Long, TX	
Issue Date/Time: Jan. 15, 2016 9:00 AM	Reviewed/Revision Date: Feb. 15, 2016	Page: 3

DELI POTATO SALAD – Flow Diagram

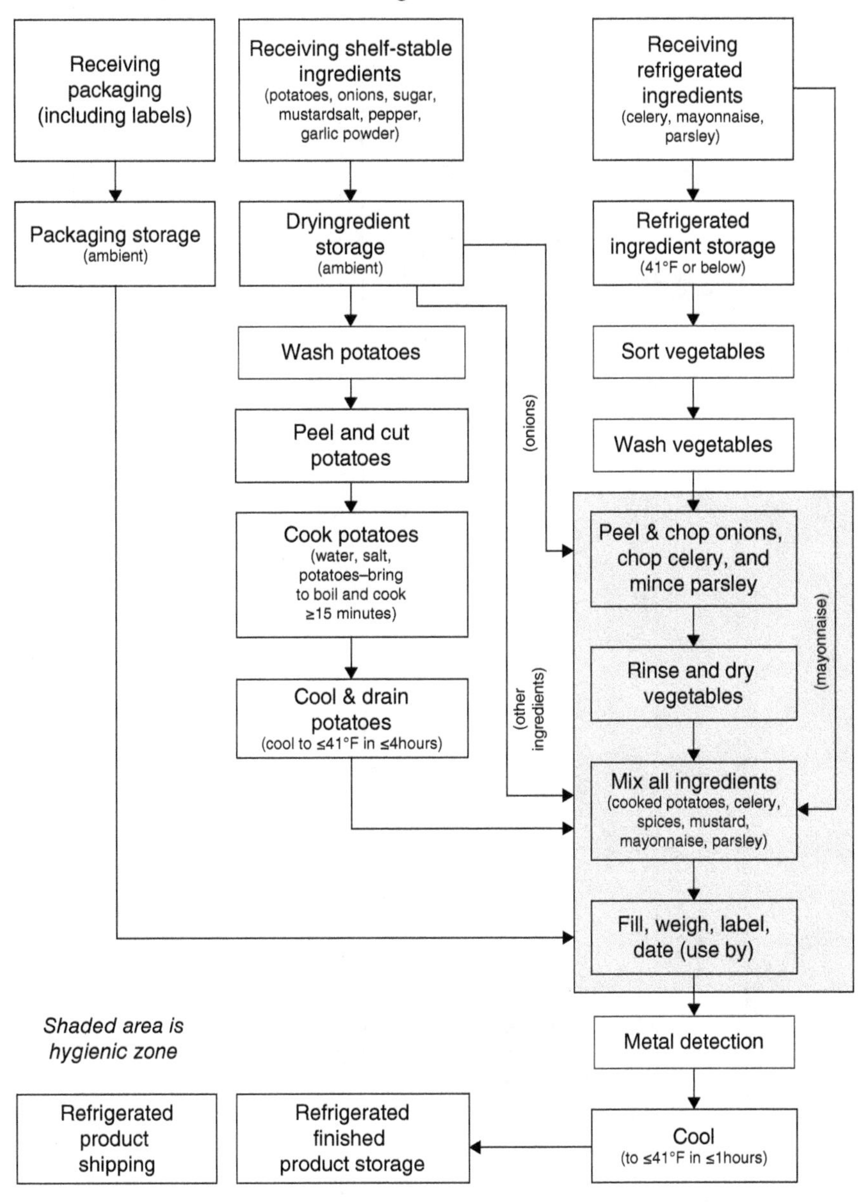

Approved By: I.M. Dave *IMD*	Reviewed By: Josh Handy *JH*	
Product(s) Code: DPS 505	Plant Name/Address: Dave's Deli Delights 5 Broad St., Long, TX	
Issue Date/Time: Jan. 15, 2016 9:00 AM	Reviewed/Revision Date: Feb. 15, 2016	Page: 4

Hazard Analysis Worksheet – Ingredients, Raw Materials, and Packaging (Part 1 of 2)
Product: DELI POTATO SALAD

(1)	(2)		(3)		(4)	(5)	(6)	
Ingredient, raw material, or packaging at receiving, storage, or use	**Identify *potential* food safety hazards – introduced, controlled, enhanced, or intentionally introduced for economic gain Biological, Chemical (including radiological), or Physical**		**Do any *potential* food safety hazards that are likely to occur require a preventive control be applied?**		**Why? Justify your decision for column 3**	**What preventive control(s) can be applied to prevent, eliminate, or significantly minimize the hazard? *E.G.,* Process Control (including CCPs), Allergen Control, Sanitation Control, Supplier Control, or other Preventive Control**	**Is the preventive control applied at this step?**	
			Yes*	No			Yes	No
Receiving potatoes	B?	Vegetative pathogens such as *Salmonella, Listeria monocytogenes, and Staphylococcus aureus*		X	*Salmonella, L. monocytogenes, and/or Staph. aureus* may be present on potatoes at a very low frequency, and subsequent cooking is adequate to destroy them.			
	C?	None						
	P?	None						

Approved By: I.M. Dave *IMD*	Reviewed By: Josh Handy *JH*
Product(s) Code: DPS 505	Plant Name/Address: Dave's Deli Delights 5 Broad St. Long, TX

Receiving other vegetables (onions, celery, and parsley)	B?	Vegetative pathogens such as *L. monocytogenes, Salmonella and Staph. aureus*	X		Suppliers of vegetables are on an approved list, and they conform to good agricultural practices (GAP) as described in the FSMA Produce Safety Rule.	***Supplier Control*** – COC (certificate of conformance) with FSMA Produce Rule for growth and harvest of all vegetables with each lot	X	
	C?	None						
	P?	None						
Receiving dry ingredients (sugar, salt, pepper, and garlic powder)	B?	Vegetative bacteria such as *Salmonella* and pathogenic *E. coli* could be present in pepper and/or garlic powder	X		*Salmonella* and pathogenic *E. coli* recalls and outbreak history.	***Supplier Control***: Verification of approved supplier's Certificates of Analysis (COA) for no *Salmonella and no pathogenic E. coli*	X	
	C?	none						
	P?	Unavoidable foreign material is associated with the growing and harvesting of spices.	X		Stones, wood, metal fragments can be collected with spices. Suppliers must use screens, magnets, etc. to remove foreign objects.	***Supplier Control***: Verification of approved supplier's Certificates of Analysis (COA) for no foreign objects		

(*continued*)

Approved By: I.M. Dave *IMD*	Reviewed By: Josh Handy *JH*	
Product(s) Code: DPS 505	Plant Name/Address: Dave's Deli Delights 5 Broad St., Long, TX	
Issue Date/Time: Jan. 15, 2016 9:00 AM	Reviewed/Revision Date: Feb. 15, 2016	Page: 6

(1)	(2)		(3)		(4)	(5)	(6)	
Ingredient, raw material, or packaging at receiving, storage, or use	**Identify *potential* food safety hazards – introduced, controlled, enhanced, or intentionally introduced for economic gain Biological, Chemical (including radiological), or Physical**		**Do any *potential* food safety hazards that are likely to occur require a preventive control be applied?**		**Why? Justify your decision for column 3**	**What preventive control(s) can be applied to prevent, eliminate, or significantly minimize the hazard? *E.G.,* Process Control (including CCPs), Allergen Control, Sanitation Control, Supplier Control, or other Preventive Control**	**Is the preventive control applied at this step?**	
Receiving mayonnaise	B?	None						
	C?	Egg allergens in mayonnaise	X		Mayonnaise is a source of egg allergens and requires labeling	***Allergen Control*** for allergen labeling at a later step		X
	P?	None						
Receiving mustard	B?	None						
	C?	None						
	P?	None						
Receiving packaging (cups, lids and labels)	B?	None						
	C?	Undeclared allergens	X		Labels must declare allergens when present, and printing errors can happen.	***Allergen Control*** for allergen labeling – label review upon receipt	X	
	P?	None						

Approved By: I.M. Dave *IMD*	Reviewed By: Josh Handy *JH*
Product(s) Code: DPS 505	Plant Name/Address: Dave's Deli Delights 5 Broad St., Long, TX

Storing shelf stable ingredients	B?	None						
	C?	None						
	P?	None						
Storing vegetables needing refrigeration and storing mayonnaise	B?	Cross-contamination of vegetables		X	Vegetables are stored in sanitized, closed containers in this refrigerated storage area (≤41°F) to avoid cross-contamination. Area and containers are cleaned and sanitized per GMPs.			
	C?	Mayonnaise has egg allergens		X	Mayonnaise is stored in a sealed container in a specific area for allergens. Good GMPs, and color coding is used to identify allergens.			
	P?	None						

(*continued*)

(1)	(2)		(3)		(4)	(5)	(6)	
Ingredient, raw material, or packaging at receiving, storage, or use	**Identify *potential* food safety hazards – introduced, controlled, enhanced, or intentionally introduced for economic gain Biological, Chemical (including radiological), or Physical**		**Do any *potential* food safety hazards that are likely to occur require a preventive control be applied?**		**Why? Justify your decision for column 3**	**What preventive control(s) can be applied to prevent, eliminate, or significantly minimize the hazard? *E.G.,* Process Control (including CCPs), Allergen Control, Sanitation Control, Supplier Control, or other Preventive Control**	**Is the preventive control applied at this step?**	
Storing packaging materials and labels	B?	None						
	C?	Undeclared allergens	X		Labels where allergens are declared must be separated from other labels for products when no allergens are present. Allergen labels are in a color-coded area.	***Allergen Control*** for allergen labeling – part of Allergen Control Program	X	
	P?	None						

* If YES, you must identify a preventive control in column 5.

Approved By: I.M. Dave *IMD*	Reviewed By: Josh Handy *JH*
Product(s) Code: DPS 505	Plant Name/Address: Dave's Deli Delights 5 Broad St., Long, TX

Hazard Analysis Worksheet – Processing Steps and Product Handling (Part 2 of 2)
Product: DELI POTATO SALAD

(1)	(2)		(3)		(4)	(5)	(6)	
Processing step or product handling. *E.G.* heating, cooling, refrigeration, freezing, pH control, zoning, staging, rework, storage, distribution, etc.	**Identify *potential* food safety hazards – introduced, controlled, enhanced, or intentionally introduced for economic gain Biological, Chemical (including radiological) or Physical**		**Do any *potential* food safety hazards that are likely to occur require a preventive control to be applied?**		**Why? Justify your decision for column 3**	**What preventive control(s) can be applied to prevent, eliminate, or significantly minimize the hazard? *E.G.,* Process Control (including CCPs), Allergen Control, Sanitation Control, Supplier Control, or other Preventive Control**	**Is the preventive control applied at this step?**	
			Yes*	No			Yes	No
Wash, peel and cut potatoes	B?	None			.			
	C?	None						
	P?	None						
Cook potatoes	B?	Vegetative pathogens such as *E. coli* and *Salmonella*		X	Validation data demonstrated that for potatoes, internal temperature and time achieved during the cooking process to achieve a palatable texture exceed parameters needed to destroy vegetative pathogens.			
	C?	None						
	P?	None						

(continued)

Approved By: I.M. Dave *IMD*	Reviewed By: Josh Handy *JH*	
Product(s) Code: DPS 505	Plant Name/Address: Dave's Deli Delights 5 Broad St., Long, TX	
Issue Date/Time: Jan. 15, 2016 9:00 AM	Reviewed/Revision Date: Feb. 15, 2016	Page: 10

(1)	(2)		(3)		(4)	(5)	(6)	
Processing step or product handling. *E.G.* heating, cooling, refrigeration, freezing, pH control, zoning, staging, rework, storage, distribution, etc.	**Identify *potential* food safety hazards – introduced, controlled, enhanced, or intentionally introduced for economic gain <u>B</u>iological, <u>C</u>hemical (including radiological) or <u>P</u>hysical**		**Do any *potential* food safety hazards that are likely to occur require a preventive control to be applied?**		**Why? Justify your decision for column 3**	**What preventive control(s) can be applied to prevent, eliminate, or significantly minimize the hazard? *E.G.,* Process Control (including CCPs), Allergen Control, Sanitation Control, Supplier Control, or other Preventive Control**	**Is the preventive control applied at this step?**	
Cool and drain potatoes	B?	None						
	C?	None						
	P?	None						
Move operations into hygienic zone area for pealing & cutting and mixing vegetables	B?	Vegetative pathogens such as *Listeria monocytogenes and Staphylococcus aureus*	X		For this Ready-to-Eat (RTE) product, care must be taken to not allow cross-contamination from workers, equipment, or the environment.	***Sanitation Control*** – Sanitation Control Program for *LM* and other pathogens includes workers (gowned and gloved) as well as equipment sanitation, hygienic zoning (segregated operations), and environmental monitoring.	X	
	C?	None						
	P?	None						
Weigh and mix ingredients in mixer	B?	None						
	C?	None						
	P?	None						
Fill cups, weigh, lid, and label	B?	None						
	C?	Undeclared Allergens on label – eggs	X		Product contains egg allergens from mayonnaise	***Allergen Control*** – declaration of allergens on label	X	
	P?	None						

Approved By: I.M. Dave *IMD*	Reviewed By: Josh Handy *JH*
Product(s) Code: DPS 505	Plant Name/Address: Dave's Deli Delights 5 Broad St., Long, TX

Metal detection	B?	None						
	C?	None						
	P?	Metal can cause injury	X		Metal fragments can come from raw materials or processing equipment.	***Process Control*** – metal detection	X	
Cool finished product in package	B?	Pathogens –*Staphylococcus aureus*	X		If present, improper cooling can allow pathogens to grow	***Process Control*** – refrigeration, Cool to ≤41°F in ≤1 hour.	X	
	C?							
	P?							
Refrigerated storage of finished product	B?	Pathogens – *Listeria monocytogenes*	X		Improper storage temperatures or temperature abuse can allow pathogens to grow.	***Process Control*** – refrigeration, Hold at ≤41°F	X	
	C?	None						
	P?	None						
Refrigerated shipping of finished product	B?	Pathogens- *Listeria monocytogenes*	X		Improper shipping temperatures or temperature abuse can allow pathogens to grow	***Process Control*** – refrigeration, Ship at ≤ 41°F	X	
	C?	None						
	P?	None						

* If YES, you must identify a preventive control in column 5.

Approved By: I.M. Dave *IMD*	Reviewed By: Josh Handy *JH*	
Product(s) Code: DPS 505	Plant Name/Address: Dave's Deli Delights 5 Broad St., Long, TX	
Issue Date/Time: Jan. 15, 2016 9:00 AM	Reviewed/Revision Date: Feb. 15, 2016	Page: 12

FOOD SAFETY PLAN SUMMARY: ALLERGEN CONTROLS, SUPPLIER CONTROLS, AND OTHER CONTROLS
PRODUCT: DELI POTATO SALAD

Hazard(s) Identified in the Hazard Analysis to Be Addressed and the (Preventive Control Measure Applied)	Performance Criteria (may include limits)	Monitoring (what, how & frequency)	Corrective Action	Verification Activities	Record-Keeping
Egg allergen in mayonnaise (*Allergen Preventive Control Program at Receiving, storing and at labeling*)	Correct labels received and stored separately from non-allergen labels Mayonnaise stored in proper location and color-coded Correct labels applied to finished product	Visual inspection of labels upon receipt. Proper storage upon receipt of labels and mayonnaise containers. Visual inspection of each finished product sample at end of run to ensure proper allergen labeling information on label.	If wrong labels are received, reject shipment. Notify supplier. If labels or mayonnaise containers are out of place, correct and notify supervisor. If label is incorrect, the product is held in quarantine until a decision is made regarding its relabeling or its disposal.	QA manager does independent check to determine labeling is correct. QA manager checks storage area. QA manager reviews and initials records daily and compares results with past testing to determine any trends.	Monitoring records Verification records Corrective action log Training records for Q.I.s
Pathogens in vegetables (*Supplier Preventive Control Program for pathogens*)	Supplier must use proper growing and harvesting processes as described in a certificate of conformance (COC) to meet FSMA Produce Rules.	Receiving clerk requests and inspects COC from qualified supplier upon each lot shipment.	If COC is missing, the load of vegetables are rejected.	QA manager sends a sample once a month to independent lab for pathogen testing. QA manager reviews receiving records weekly. Audit of supplier- conducted annually and reviewed by QA manager.	Receiving records Verification records from QA manager and lab results Corrective action log

Approved By: I.M. Dave *IMD*	Reviewed By: Josh Handy *JH*
Product(s) Code: DPS 505	Plant Name/Address: Dave's Deli Delights 5 Broad St., Long, TX

Pathogens and foreign objects in dry ingredients (*Supplier Preventive Control Program for pathogens and foreign objects*)	Suppliers must provide testing results for absence of *Salmonella* and pathogenic STEC *E. coli* as described in a certificate of analysis (COA). COA also includes test results for foreign objects.	Receiving clerk requests and inspects COA from qualified supplier upon each ship-ment looking at test results.	If COA is missing or out of specification, the load of dry ingredient is rejected.	QA manager sends a dry ingredient sample once a month to independent lab for pathogen testing. QA manager reviews receiving records weekly. QA manager reviews consumer complaints for identification of foreign objects. Audit of supplier-conducted annually and reviewed by QA manager.	Receiving records Verification records from QA manager and lab results Corrective action log
Egg allergen in Mayonnaise (Allergen Preventive Control program at labeling)	Correct labels applied to finished product.	Visual inspection by line operator of each label placed in packaging machine and of label on finished product at end of run to ensure proper allergen labeling information on label.	If label is incorrect, the product is held in quarantine until a decision is made regarding its relabeling or its disposal.	QA manager does inde-pendent check to deter-mine labeling is correct. QA manager reviews and initials records daily and compares results with past testing to determine any trends.	Monitoring records Verification records Corrective action log

FOOD SAFETY PLAN SUMMARY: PROCESS CONTROLS AND SANITATION CONTROLS
PRODUCT: DELI POTATO SALAD

Preventive Controls: Critical Control Point (CCP) or Sanitation Preventive Control	Hazard(s) to Be Addressed in Food Safety Plan	Critical Limits for Each Control Measure	Monitoring				Corrective Action	Verification Activities	Record-keeping Procedures
			What	How	Frequency	Who			
Sanitation preventive Control (Hygienic Zoning in weighing, mixing and packaging area)	Cross-contamination of product by pathogens (*Salmonella* and/or *Listeria spp.*) in mixing area	Food Contact equipment must be properly cleaned and sanitized Non-food contact surfaces cleaned and sanitized Line workers handling food are wearing gowns and gloves	Food contact surfaces Floors, walls, and drains Line workers protective clothing and gloves	Sponge swabs collected from chopping, weighing, processing equipment. Swabs from drains. All composite samples are prepared and sent to lab for testing Sign-in suit-up log when gowned and gloved	Every batch at end of shift after equipment has been used Daily at end of shift after equipment has been used Once every batch	Line workers that prepare and package product Line workers that clean and sanitize equipment and hygienic zone area Line workers use suit-up log and initial log	If food-contact test results are positive, use steps in SOP (XYZ) to further identify problem, such as individual site sampling; identify root cause; and determine if lot needs destruction For non-food positive, refer to SOP (ABC) and conduct intensive cleaning of area and equipment Retrain as needed	Environmental monitoring of food contact surfaces and non-food contact areas (e.g., drains) to verify effective sanitation Visual inspection of workers clothing/gloves once during shift by QA manger	Sampling and testing log Verification records of sanitizing procedures Corrective action records Training records for Q.I.s
Process preventive control (Metal detection)	Metal inclusion may cause dental injury or other physical injury	Metal detector present and operating correctly	All the product passes through an operating metal detector	Visual exam-ination that metal detector is “on” and reject device is working on test wand	Beginning middle and end of shift	Line worker	If the product is processed without metal detection, hold it for metal detection. Correct operating procedures to ensure that the product is not processed without metal detection. If metal is found in product, segregate product, inspect back to the last good check, rework, or discard product..	Pass X mm ferrous and Y mm nonferrous and stainless standard wands through detector at start-up, middle, and end of shift to assure equipment is functioning. Weekly review of Metal Detector Log and Corrective Action and Verification. Annual calibration by detector’s manufacturer.	Metal detector log Manufacturer’ s validation study that determined detector setting and sensitivity standards Corrective action records Q.I’s training records

Approved By: I.M. Dave *IMD*	Reviewed By: Josh Handy *JH*
Product(s) Code: DPS 505	Plant Name/Address: Dave’s Deli Delights 5 Broad St., Long, TX

Process preventive control (cooling of packaged product)	Growth and toxin production of vegetative pathogens such as *Staphylococcus aureus and L. monocytogenes* can occur if product is not cooled properly	Product temperature during cooling must reach ≤41°F in 60 minutes or less.	Potato Salad temperature	Recording thermo-meter that is placed directly in center of potato salad sample	Every batch at end of the cooling cycle.	Line operator	If temperature limit is exceeded, then, 1) segregate product, evaluate product, or discard as appropriate; 2) identify root cause of temperature problem; 3) conduct training to prevent recurrence.	Daily review of temperature cooling log by Production Supervisor Accuracy of thermometer checked weekly by QA Tech. Quarterly thermometer calibration	Cooling temperature log Thermometer calibration records Verification records Validation information describing critical refrigeration parameters Corrective action records Training records for Q.I.s
Process preventive control (Refrigeration in storage and shipping)	Growth and toxin production of vegetative pathogens such as *Salmonella, E. coli, Staphylococcus. aureus, and L. monocytogenes* can occur if product is temperature abused.	Product temperature during storage and shipping must not exceed 41°F.	Potato Salad storage temperature	Recording thermo-meter that is placed directly in center of potato salad sample	Every new batch at end of shift and same batch next day	Shipping clerk	If temperature limit is exceeded, then, 1) segregate product, evaluate product, or discard as appropriate; 2) identify root cause of temperature failure; 3) conduct training to prevent recurrence	Daily review of temperature storage and shipping log by Supervisor Accuracy of thermometer checked weekly by QA Tech. Quarterly thermometer calibration	Storage log and shipping log Thermometer calibration records Verification records Validation report for determining critical refrigeration parameters Corrective action records Training records for Q.I.s

Approved By: I.M. Dave *IMD*	Reviewed By: Josh Handy *JH*	
Product(s) Code: DPS 505	Plant Name/Address: Dave's Deli Delights 5 Broad St., Long, TX	
Issue Date/Time: Jan. 15, 2016 9:00 AM	Reviewed/Revision Date: Feb. 15, 2016	Page: 4

Macaroni & Cheese Frozen Meal – Example Food Safety Plan

Company Overview

This example company (Mom's Macaroni Inc.) is a mid-sized firm that makes a variety of frozen entrees that are intended to be cooked prior to consumption. Products include Macaroni and Cheese, Lasagna and Spaghetti. Products are made 5 days a week in one 8-hour production shift, followed by 4 hours for sanitation. The plant follows Good Manufacturing Procedures (GMPs) regulations as described in 21 CFR Part 117, Subpart B. Prerequisite programs (PRP) for cleaning and sanitizing, as well as other PRPs, are written and performed by trained workers as standard operating procedures (SOPs). Workers record the results of their tasks when the SOP is completed.

This Food Safety Plan covers production of Macaroni and Cheese. The pasta cooking equipment is shared to make other products. The company also has a written recall plan.

Product Description/ Product Name	Macaroni and Cheese Frozen Meal
Product Description (including any important food safety characteristics)	Macaroni and Cheese is a frozen, ready-to-cook meal packed in a metal tray and sealed with a heat-sealable film. The frozen product is placed in a box and warehoused, distributed, and retailed frozen.
Ingredients and Raw Materials	Cooked macaroni (wheat flour), processed cheese sauce, milk powder, sodium phosphate, citric acid, yellow 5, and salt. Potable water is treated and tested per EPA requirements by the city.
Packaging Used	8" aluminum tray with heat-sealable film inside pre-printed carton. 10 ounces serving size.
Intended Use	Fully cook before serving (NRTE – not ready to eat)
Intended Consumers	General public
Shelf Life	1 to 2 years frozen
Labeling Instructions	Keep frozen. Oven cooking instructions. Allergen labeling.
Storage and Distribution	Frozen storage and retail distribution

Approved By: B.G. Boss *BGB*	Reviewed By: I.M. Right *IMR*	
Product(s) Code: MC 1001	Plant Name/Address: Mom's Mac 123 Main St. Atlanta, GA	
Issue Date/Time: Jan. 1, 2016 8:00 AM	Reviewed/ Revision Date: Jan. 30, 2016	Page: 1

This generic Food Safety Plan was developed for example purposes only and is not intended to replace a processor's hazard analysis or the development of a facility's own Food Safety Plan.

Process Narrative

Receiving Ingredients: Ingredients and raw materials are purchased from domestic suppliers complying with recognized food safety and quality schemes. Ingredients are stored according to manufacturers' recommendations.

Receiving Packaging: Pre-labeled cartons and 10 ounce aluminum trays with heat-sealable lids are received in bulk. Labeled cartons are reviewed for conformance with product allergen requirements and ingredients. Specifications require food-grade material compatible for frozen food products.

Receiving Shelf-stable Ingredients:

- Cooked macaroni (wheat flour): Received dry from sole-source supplier.
- Milk powder: Received in bag-in-boxes (50 pounds).
- Sodium phosphate, citric acid, and yellow 5: Received from U.S.-approved supplier.
- Salt: Received in bags from sole-source broker. Specifications require food grade salt.

Receiving Refrigerated Ingredients:

- Pasteurized processed cheese sauce: Received in 25-gallon plastic bags in storage boxes from an approved supplier.

Storing Ingredients and Packaging:

- **Packaging storage:** Cartons, trays, cases, and lid materials are stored in the dry storage room in the packaging area, arranged by product code to avoid mixing of packaging. Packaging is used first-in-first-out, and partially used shipping containers are closed during storage.
- **Dry ingredient storage:** Macaroni, milk powder, chemical ingredients, and salt are stored in the dry storage room in the ingredient area, arranged by ingredient code number. All containers are sealed to avoid food allergen cross-contact and cross-contamination during storage. Ingredients containing food allergens (wheat and milk powder) are identified and stored in specific locations with similar allergenic ingredients.
- **Refrigerated ingredient storage:** Cheese sauce is stored in a cooler that is kept at (≤41°F) and used on a first-in-first-out basis.

Cook Macaroni: Salt is added to potable water in a steam-jacketed kettle and is brought to boiling. The macaroni is added to boiling water and cooked for at least 15 minutes to achieve acceptable texture. Product temperature exceeds 165°F in this processing time. Only macaroni is processed in the cook area when this product is made. The kettle is drained, cooked macaroni is retained, and water discarded.

Cool, Drain Macaroni: Potable cold water is used to cool the drained macaroni in <6 hour to 41°F. Cold water temperatures are 50–55°F depending on the season. Macaroni is cooled by running cold water through it or adding potable ice. Generally, the macaroni is brought to the cold water temperature in <45 minutes, after which it is drained and transferred to the assembly area, where it is mixed with other ingredients. A batch of drained macaroni is assembled with other ingredients and packaged in <30 minutes after delivery to the assembly area. Any unused macaroni is saved in refrigerated storage (≤41°F) for no more than 48 hours. It is generally used in the next batch as rework.

Mix All Ingredients: Dry ingredients are blended in a heated kettle-type ribbon mixer with cheese sauce and milk powder to distribute them just prior to adding cooked macaroni. They are blended to a uniform mixture. A batch is blended in <15 minutes. The blended ingredient temperature is typically above ambient after mixing.

Fill, Weigh, Lid: 10 oz. trays are filled, weighed, and lidded.

Freezing: Packaged product passes through freezer tunnel. Product temperature is ≤5°F in ≤60 minutes.

Cartoning and Casing: Frozen packages are placed in pre-labeled cardboard cartons and are boxed in cases – 24 per case. Carton has ingredient and allergen information that declares the wheat and milk allergens. Labels on cartons are matched to the specific description and product number. Oven cooking directions are provided on the label.

Frozen Finished Product Storage: Finished product is transferred to frozen storage (≤0°F).

Frozen Finished Product Shipping: Product is shipped in freezer trucks to customers (convenience stores, office cafeterias, quick serve restaurants, and grocery stores) under frozen conditions (≤0°F).

Approved By: B.G. Boss *BGB*	Reviewed By: I.M. Right *IMR*	
Product(s) Code: MC 1001	Plant Name/Address: Mom's Mac 123 Main St. Atlanta, GA	
Issue Date/Time: Jan. 1, 2016 8:00 AM	Reviewed/ Revision Date: Jan. 30, 2016	Page: 3

Macaroni and Cheese Frozen Meal – Flow Diagram

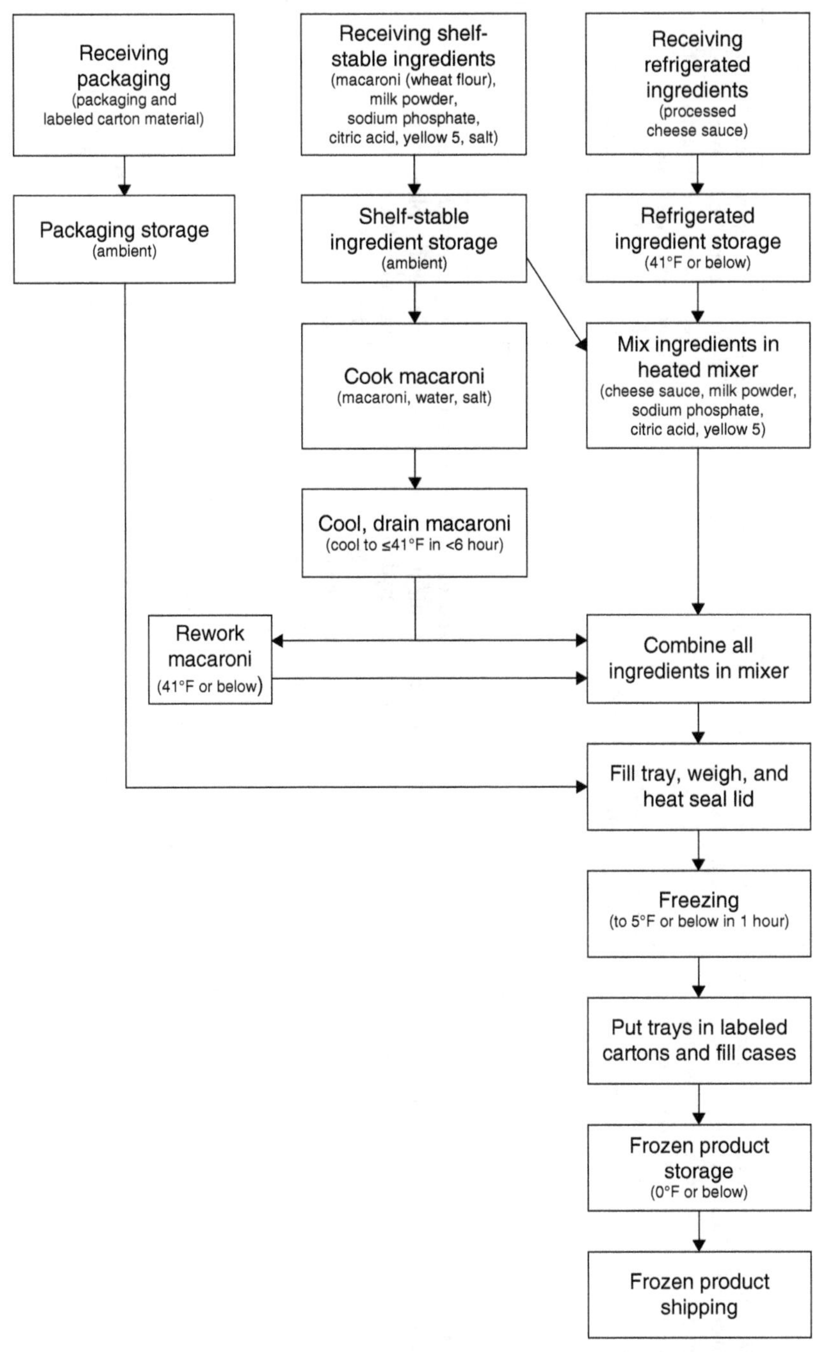

Approved By: B.G. Boss *BGB*	Reviewed By: I.M. Right *IMR*	
Product(s) Code: MC 1001	Plant Name/Address: Mom's Mac 123 Main St. Atlanta, GA	
Issue Date/Time: Jan. 1, 2016 8:00 AM	Reviewed/ Revision Date: Jan. 30, 2016	Page: 4

Hazard Analysis Worksheet – Ingredients, Raw Materials, and Packaging (Part 1 of 2)
Product: Macaroni and Cheese Meal

(1)	(2)		(3)		(4)	(5)	(6)	
Ingredient, raw material or packaging at: receiving, storage or use	Identify *potential* food safety hazards – introduced, controlled, enhanced or intentionally introduced for economic gain Biological, Chemical (including radiological), or Physical		Do any *potential* food safety hazards that are likely to occur require a preventive control to be applied?		Why? Justify your decision for column 3	What preventive control(s) can be applied to prevent, eliminate, or significantly minimize the hazard? *E.G.*, Process Control (including CCPs), Allergen Control, Sanitation Control, Supplier Control or other Preventive Control	Is the preventive control applied at this step?	
			Yes*	No			Yes	No
Receiving macaroni	B?	Spore-forming pathogens such as *B. cereus*	X		*B. cereus* spores may be present in dry macaroni, and outbreaks due to growth after hydration have occurred in rice & pasta. Levels present at receiving are not hazardous and will not change as long as the macaroni is dry.	Subsequent ***Process Control*** cooling step prevents *B. cereus* growth and toxin formation in rehydrated macaroni.		X
		Vegetative pathogens such as *Salmonella*		X	*Salmonella* may be present in macaroni at a very low frequency, and subsequent cooking is more than adequate to destroy it.			
	C?	Wheat allergens	X		Macaroni is made from wheat flour. Wheat is a common food allergen and needs proper labeling	***Allergen Control*** for allergen labeling at later steps		X
	P?	None						

Approved By: B.G. Boss *BGB*	Reviewed By: I.M. Right *IMR*
Product(s) Code: MC 1001	Plant Name/Address: Mom's Mac 123 Main St. Atlanta, GA

Receiving pasteurized cheese sauce	B?	Vegetative pathogens such as *L. monocytogenes* and *Salmonella*	X		Supplier gives a validated heat pasteurization treatment to destroy vegetative pathogens	***Supplier Control*** – COC (certificate of conformance for all cheese sauce made using an approved pasteurization process) with each lot	X	
	C?	Milk allergens	X		Cheese is a source of milk allergens and requires labeling	***Allergen Control*** for allergen labeling at a later step		X
	P?	None						
Receiving milk powder	B?	Vegetative pathogens such as *Salmonella*	X		Supplier provides a COA (certificate of analysis) for each lot to show *Salmonella* negative	***Supplier Control*** – COA at receiving	X	
	C?	Allergens	X		Milk powder is a source of milk allergens and requires labeling	***Allergen Control*** for allergen labeling at a later step		X
	P?	None						

(continued)

(1)	(2)		(3)		(4)	(5)	(6)	
Ingredient, raw material or packaging at: receiving, storage or use	**Identify *potential* food safety hazards – introduced, controlled, enhanced or intentionally introduced for economic gain <u>B</u>iological, <u>C</u>hemical (including radiological), or <u>P</u>hysical**		**Do any *potential* food safety hazards that are likely to occur require a preventive control to be applied?**		**Why? Justify your decision for column 3**	**What preventive control(s) can be applied to prevent, eliminate, or significantly minimize the hazard? *E.G.,* Process Control (including CCPs), Allergen Control, Sanitation Control, Supplier Control or other Preventive Control**	**Is the preventive control applied at this step?**	
Receiving packaging (includes labeled cartons)	B?	None						
	C?	Undeclared allergens	X		Labels must declare allergens when present, and printing errors can happen	***Allergen Control*** for allergen labeling – label review upon receipt	X	
	P?	None						
Storing macaroni and milk powder	B?	None						
	C?	Allergen cross-contact during storage		X	Good Manufacturing Practices (GMPs) allow only sealed containers of allergenic ingredients in dry storage, thus cross-contact is unlikely			
	P?	None						

Approved By: B.G. Boss *BGB*	Reviewed By: I.M. Right *IMR*
Product(s) Code: MC 1001	Plant Name/Address: Mom's Mac 123 Main St. Atlanta, GA

Storing cheese sauce	B?	None						
	C?	Allergen cross-contact during storage		X	Only closed containers of allergenic ingredients are stored in this refrigerated storage area to avoid cross-contact			
	P?	None						
Storing other dry ingredients including packaging	B?	None						
	C?	None						
	P?	None						

* If YES, you must identify a preventive control in column 5.

Approved By: B.G. Boss *BGB*	Reviewed By: I.M. Right *IMR*	
Product(s) Code: MC 1001	Plant Name/Address: Mom's Mac 123 Main St. Atlanta, GA	
Issue Date/Time: Jan. 1, 2016 8:00 AM	Reviewed/ Revision Date: Jan. 30, 2016	Page: 8

Hazard Analysis Worksheet – Processing Steps and Product Handling (Part 2 of 2)
Product: Macaroni and Cheese Meal

(1)	(2)		(3)		(4)	(5)	(6)	
Processing step or product handling. *E.G.*, heating, cooling, refrigeration, freezing, pH control, zoning, staging, rework, storage, distribution, etc.	**Identify *potential* food safety hazards – introduced, controlled, enhanced, or intentionally introduced for economic gain Biological, Chemical (including radiological), or Physical**		**Do any *potential* food safety hazards that are likely to occur require a preventive control to be applied?**		**Why? Justify your decision for column 3**	**What preventive control(s) can be applied to prevent, eliminate, or significantly minimize the hazard? *E.G.*, Process Control (including CCPs), Allergen Control, Sanitation Control, Supplier Control, or other Preventive Control**	**Is the preventive control applied at this step?**	
			Yes*	No			Yes	No
Cook macaroni	B?	Vegetative pathogens such as *Salmonella*		X	Validation data demonstrated that for macaroni, temperatures achieved during the cooking process (boiling water for ≥15 minutes) to achieve a palatable texture far exceed those needed to destroy vegetative pathogens.			
	C?	Allergens		X	GMP sanitation SOP between runs ensure no cross-contact in cooking kettle			
	P?	None						
Cool and drain macaroni	B?	Spore-forming pathogens (such as *B. cereus* and *C. perfringens*)	X		Spores may survive cooking. Growth and toxin production could occur during slow cooling.	***Process Control*** – Time and temperature control for cooling macaroni	X	
	C?	None						
	P?	None						

Approved By: B.G. Boss *BGB*	Reviewed By: I.M. Right *IMR*
Product(s) Code: MC 1001	Plant Name/Address: Mom's Mac 123 Main St. Atlanta, GA

Weigh and mix ingredients in ribbon blender	B?	None						
	C?	None						
	P?	None						
Fill trays, weigh and lid	B?	None						
	C?	None						
	P?	None						
Freeze sealed trays	B?	None						
	C?	None						
	P?	None						
Insert in labeled carton	B?	None						
	C?	Undeclared Allergens –wheat and milk	X		Product contains wheat and milk allergens	***Allergen Control*** – declaration allergens on label	X	
	P?	None						
Store finished product	B?	None						
	C?	None						
	P?	None						
Ship finished product	B?	None						
	C?	None						
	P?	None						

* If YES, you must identify a preventive control in column 5.

Approved By: B.G. Boss *BGB*	Reviewed By: I.M. Right *IMR*
Product(s) Code: MC 1001	Plant Name/Address: Mom's Mac 123 Main St. Atlanta, GA
Issue Date/Time: Jan. 1, 2016 8:00 AM	Reviewed/ Revision Date: Jan. 30, 2016 — Page: 10

FOOD SAFETY PLAN SUMMARY: ALLERGEN CONTROLS, SUPPLIER CONTROLS AND OTHER CONTROLS
PRODUCT: Macaroni and Cheese Meal

Hazard(s) Identified in the Hazard Analysis to be Addressed and the (Preventive Control Measure Applied)	Performance Criteria (may include limits)	Monitoring (what, how & frequency)	Corrective Action	Verification Activities	Record-Keeping
Allergen in macaroni (wheat flour), cheese sauce (milk) and milk powder *(Allergen Preventive Control Program)*	Correct labels received and correct labels applied to finished product	Visual inspection of labels upon receipt. Visual inspection by line operator of labels in packaging machine before and at end of run to ensure proper allergen labeling information on label.	If wrong labels are received, reject shipment. Notify supplier. If label is incorrect, the product is held in quarantine until a decision is made regarding its relabeling or its disposal.	QA manager does independent check to determine labeling is correct. A sample labeled product is taken for verification of proper label. QA manager reviews and initials records daily and compares results with past testing to determine any trends.	Monitoring records Verification records Corrective action log Training records for Q.I.s

Approved By: B.G. Boss *BGB*	Reviewed By: I.M. Right *IMR*
Product(s) Code: MC 1001	Plant Name/Address: Mom's Mac 123 Main St. Atlanta, GA

Pathogens in cheese sauce (*Supplier Preventive Control Program for pasteurization*)	Supplier must use validated pasteurization process as described in a certificate of conformance (COC) to the process	Receiving clerk requests and inspects COC from qualified supplier upon each lot shipment	If COC is missing, the load of cheese sauce is rejected	QA manager sends a sample once a month to independent lab for pathogen testing QA manager reviews receiving records weekly. Audit of supplier – conducted annually and reviewed by QA manager	Receiving records Verification records from QA manager and lab results Corrective action log
Pathogens in milk powder (*Supplier Preventive Control Program for* Salmonella *specification*)	Supplier must provide milk powder absent of *Salmonella* as described in a certificate of analysis (COA)	Receiving clerk requests and inspects COA from qualified supplier upon each shipment.	If COA is missing or out of specification, the load of milk powder is rejected.	QA manager sends a milk powder sample once a month to independent lab for pathogen testing. QA manager reviews receiving records weekly. Audit of supplier- conducted annually and reviewed by QA manager.	Receiving records Verification records from QA manager and lab results Corrective action log

FOOD SAFETY PLAN SUMMARY: PROCESS CONTROLS AND SANITATION CONTROLS
PRODUCT: Macaroni and Cheese Meal

Preventive Controls: Critical Control Point (CCP) or Sanitation Preventive Control	Hazard(s) to be Addressed in Food Safety Plan	Critical Limits for Each Control Measure	Monitoring				Corrective Action	Verification Activities	Record-keeping Procedures
			What	How	Frequency	Who			
Cool macaroni (Process preventive control)	Growth and toxin production of spore-formers such as *Clostridium perfringens* and *B. cereus*	Chill macaroni within 2 hours to 70°F and within a total of 6 hours to 41°F or less.	Macaroni temperature and time	Recording thermo-meter that is placed directly in macaroni	During every batch- check temp once every 20 min, so that process adjustment (e.g., add more cold water or ice) can be made before a critical limit is exceeded.	Line Cook	If temperature is not achieved in the specified time, then 1) add more cold water or ice to achieve tempera-ture; 2) segregate product, evaluate product, rework, or discard as appropriate; 3) identify root cause; 4) conduct training to prevent recurrence.	Daily review of macaroni cooling log by Supervisor Accuracy of thermometer checked weekly by QA Tech. Quarterly thermometer calibration	Macaroni cooling log Thermometer calibration records Verification records Validation report for determining critical limit cooling parameters Corrective action records Training records for Q.I.s

Approved By: B.G. Boss *BGB*	Reviewed By: I.M. Right *IMR*
Product(s) Code: MC 1001	Plant Name/Address: Mom's Mac 123 Main St. Atlanta, GA

9

FSMA Regulations: cGMPs, Hazard Analysis, and Risk-Based Preventive Controls for Human Foods

FDA's final regulations on Current Good Manufacturing Practice, Hazard Analysis and Risk-based Controls for Human Foods was published on September 17, 2015. Like the proposed rules, the final regulation is focused on a preventive approach to food safety. Following the publication of the final rules, compliance dates were established for Current Good Manufacturing Practices (cGMPs) and preventive controls, other than supplier controls. These dates, from the date of publication (September 17, 2015), were:

- Very small businesses (< $1,000,000 annual food sales) – three years (except for records to support its status as a very small business – January 1, 2016)
- Businesses subject to the Pasteurized Milk Ordinance (PMO) – three years
- Small businesses (<500 full-time employees) – two years
- All other businesses – one year
- Compliance dates for the written assurances in the customer provisions in part 117.136 and related rules – two years

Compliance dates, after publication of the final rule, for the requirements of the supply chain program:

- Receiving facility is a small business, and its supplier will not be subject to the human preventive controls rule or the produce safety rule: two years.
- Receiving facility is a small business, and its supplier will be subject to the human preventive controls rule or the produce safety rule: two years or six months after the supplier is required to comply with the applicable rule, whichever is later.
- Receiving facility is not a small or very small business, and its supplier will not be subject to the human preventive controls rule or the produce safety rule: eighteen months.
- Receiving facility is not a small or very small business, and its supplier will be subject to the human preventive controls rule or the produce safety rule: six months after the supplier is required to comply with the applicable rule.

FSMA and Food Safety Systems: Understanding and Implementing the Rules,
First Edition. Jeffrey T. Barach.

FDA Regulations on cGMP's, Hazard Analysis and Risk-based Preventive Controls for Human Foods

21 CFR Part 117: Current Good Manufacturing Practice, Hazard Analysis and Risk-based Preventive Controls for Human Food

Part 117—Current Good Manufacturing Practice, Hazard Analysis, and Risk–Based Preventive Controls for Human Food

Subpart A—General Provisions

Subpart B—Current Good Manufacturing Practice

Subpart C—Hazard Analysis and Risk-Based Preventive Controls

117.136 Circumstances in which the owner, operator, or agent in charge of a manufacturing/processing facility is not required to implement a preventive control.
117.137 Provision of assurances required under § 117.136(a)(2), (3), and (4).
117.139 Recall plan.
117.140 Preventive control management components.
117.145 Monitoring.
117.150 Corrective actions and corrections.
117.155 Verification.
117.160 Validation.
117.165 Verification of implementation and effectiveness.
117.170 Reanalysis.
117.180 Requirements applicable to a preventive controls qualified individual and a qualified auditor.
117.190 Implementation records required for this subpart.

Subpart D—Modified Requirements

117.201 Modified requirements that apply to a qualified facility.
117.206 Modified requirements that apply to a facility solely engaged in the storage of unexposed packaged food.

Subpart E—Withdrawal of a Qualified Facility Exemption

117.251 Circumstances that may lead FDA to withdraw a qualified facility exemption.
117.254 Issuance of an order to withdraw a qualified facility exemption.
117.257 Contents of an order to withdraw a qualified facility exemption.
117.260 Compliance with, or appeal of, an order to withdraw a qualified facility exemption.
117.264 Procedure for submitting an appeal.
117.267 Procedure for requesting an informal hearing.
117.270 Requirements applicable to an informal hearing.
117.274 Presiding officer for an appeal and for an informal hearing.
117.277 Timeframe for issuing a decision on an appeal.
117.280 Revocation of an order to withdraw a qualified facility exemption.
117.284 Final agency action.
117.287 Reinstatement of a qualified facility exemption that was withdrawn.

Subpart F—Requirements Applying to Records That Must Be Established and Maintained

117.301 Records subject to the requirements of this subpart.
117.305 General requirements applying to records.
117.310 Additional requirements applying to the food safety plan.

Subpart G—Supply-Chain Program

Authority: 21 U.S.C. 331, 342, 343, 350d note, 350 g, 350 g note, 371, 374; 42 U.S.C. 243, 264, 271.

Subpart A—General Provisions

§ 117.1 Applicability and status.

a) The criteria and definitions in this part apply in determining whether a food is:
   1) Adulterated within the meaning of: or
      i) Section 402(a)(3) of the Federal Food, Drug, and Cosmetic Act in that the food has been manufactured under such conditions that it is unfit for food; or
      ii) Section 402(a)(4) of the Federal Food, Drug, and Cosmetic Act in that the food has been prepared, packed, or held under insanitary conditions whereby it may have become contaminated with filth, or whereby it may have been rendered injurious to health; and
   2) In violation of section 361 of the Public Health Service Act (42 U.S.C. 264).
b) The operation of a facility that manufactures, processes, packs, or holds food for sale in the United States if the owner, operator, or agent in charge of such facility is required to comply with, and is not in compliance with, section 418 of the Federal Food, Drug, and Cosmetic Act or subpart C, D, E, or F of this part is a prohibited act under section 301(uu) of the Federal Food, Drug, and Cosmetic Act.
c) Food covered by specific current good manufacturing practice regulations also is subject to the requirements of those regulations.

§ 117.3 Definitions.

The definitions and interpretations of terms in section 201 of the Federal Food, Drug, and Cosmetic Act apply to such terms when used in this part. The following definitions also apply:

Acid foods or *acidified foods* means foods that have an equilibrium pH of 4.6 or below.

Adequate means that which is needed to accomplish the intended purpose in keeping with good public health practice.

Affiliate means any facility that controls, is controlled by, or is under common control with another facility.

Allergen cross-contact means the unintentional incorporation of a food allergen into a food.

Audit means the systematic, independent, and documented examination (through observation, investigation, records review, discussions with employees of the audited entity, and, as appropriate, sampling and laboratory analysis) to assess a supplier's food safety processes and procedures.

Batter means a semifluid substance, usually composed of flour and other ingredients, into which principal components of food are dipped or with which they are coated, or which may be used directly to form bakery foods.

Blanching, except for tree nuts and peanuts, means a prepackaging heat treatment of foodstuffs for an adequate time and at an adequate temperature to partially or completely inactivate the naturally occurring enzymes and to effect other physical or biochemical changes in the food.

Calendar day means every day shown on the calendar.

Correction means an action to identify and correct a problem that occurred during the production of food, without other actions associated with a corrective action procedure (such as actions to reduce the likelihood that the problem will recur, evaluate all affected food for safety, and prevent affected food from entering commerce).

Critical control point means a point, step, or procedure in a food process at which control can be applied and is essential to prevent or eliminate a food safety hazard or reduce such hazard to an acceptable level.

Defect action level means a level of a non-hazardous, naturally occurring, unavoidable defect at which FDA may regard a food product "adulterated" and subject to enforcement action under section 402(a)(3) of the Federal Food, Drug, and Cosmetic Act. *Environmental pathogen* means a pathogen capable of surviving and persisting within the manufacturing, processing, packing, or holding environment such that food may be contaminated and may result in foodborne illness if that food is consumed without treatment to significantly minimize the environmental pathogen. Examples of environmental pathogens for the purposes of this part include *Listeria monocytogenes* and *Salmonella* spp. but do not include the spores of pathogenic sporeforming bacteria.

Facility means a domestic facility or a foreign facility that is required to register under section 415 of the Federal Food, Drug, and Cosmetic Act, in accordance with the requirements of part 1, subpart H of this chapter.

Farm means farm as defined in § 1.227 of this chapter.

FDA means the Food and Drug Administration.

Food means food as defined in section 201(f) of the Federal Food, Drug, and Cosmetic Act and includes raw materials and ingredients.

Food allergen means a major food allergen as defined in section 201(qq) of the Federal Food, Drug, and Cosmetic Act.

Food-contact surfaces are those surfaces that contact human food and those surfaces from which drainage, or other transfer, onto the food or onto surfaces that contact the food ordinarily occurs during the normal course of operations. "Food-contact surfaces" includes utensils and food-contact surfaces of equipment.

Full-time equivalent employee is a term used to represent the number of employees of a business entity for the purpose of determining whether the business qualifies for the small business exemption. The number of full-time equivalent employees is determined by dividing the total number of hours of salary or wages paid directly to employees of the business entity and of all of its affiliates and subsidiaries by the number of hours of work in 1 year, 2,080 hours (*i.e.*, 40 hours · 52 weeks). If the result is not a whole number, round down to the next lowest whole number.

Harvesting applies to farms and farm mixed-type facilities and means activities that are traditionally performed on farms for the purpose of removing raw agricultural commodities from the place they were grown or raised and preparing them for use as food. Harvesting is limited to activities performed on raw agricultural commodities, or on processed foods created by drying/dehydrating a raw agricultural commodity without additional manufacturing/processing, on a farm. Harvesting does not include activities that transform a raw agricultural commodity into a processed food as defined in section 201(gg) of the Federal Food, Drug, and Cosmetic Act. Examples of harvesting include cutting (or otherwise separating) the edible portion of the raw agricultural commodity from the crop plant and removing or trimming part of the raw agricultural commodity (*e.g.*, foliage, husks, roots or stems). Examples of harvesting also include cooling, field coring, filtering, gathering, hulling, removing stems and husks from, shelling, sifting, threshing, trimming of outer leaves of, and washing raw agricultural commodities grown on a farm.

Hazard means any biological, chemical (including radiological), or physical agent that has the potential to cause illness or injury.

Hazard requiring a preventive control means a known or reasonably foreseeable hazard for which a person knowledgeable about the safe manufacturing, processing, packing, or holding of food would, based on the outcome of a hazard analysis (which includes an assessment of the severity of the illness or injury if the hazard were to occur and the probability that the hazard will occur in the absence of preventive controls), establish one or more preventive controls to significantly minimize or prevent the hazard in a food and components to manage those controls (such as monitoring, corrections or corrective actions, verification, and records) as appropriate to the food, the facility, and the nature of the preventive control and its role in the facility's food safety system.

Holding means storage of food and also includes activities performed incidental to storage of a food (*e.g.*, activities performed for the safe or effective storage of that food, such as fumigating food during storage, and drying/dehydrating raw agricultural commodities when the drying/ dehydrating does not create a distinct commodity (such as drying/dehydrating hay or alfalfa)). Holding also includes activities performed as a practical necessity for the distribution of that food (such as blending of the same raw agricultural commodity and breaking down pallets), but does not include activities that transform a raw agricultural commodity into a processed food as defined in section 201(gg) of the Federal Food, Drug, and Cosmetic Act. Holding facilities could include warehouses, cold storage facilities, storage silos, grain elevators, and liquid storage tanks.

Known or reasonably foreseeable hazard means a biological, chemical (including radiological), or physical hazard that is known to be, or has the potential to be, associated with the facility or the food.

Lot means the food produced during a period of time and identified by an establishment's specific code.

Manufacturing/processing means making food from one or more ingredients, or synthesizing, preparing, treating, modifying or manipulating food, including food crops or ingredients. Examples of manufacturing/processing activities include: Baking, boiling, bottling, canning, cooking, cooling, cutting, distilling, drying/ dehydrating raw agricultural commodities to create a distinct commodity (such as drying/ dehydrating grapes to produce raisins), evaporating, eviscerating, extracting juice, formulating, freezing, grinding, homogenizing, irradiating, labeling, milling, mixing, packaging (including modified atmosphere packaging), pasteurizing, peeling, rendering, treating to manipulate ripening, trimming, washing, or waxing. For farms and farm mixed-type facilities, manufacturing/ processing does not include activities that are part of harvesting, packing, or holding.

Microorganisms means yeasts, molds, bacteria, viruses, protozoa, and microscopic parasites and includes species that are pathogens. The term "undesirable microorganisms" includes those microorganisms that are pathogens, that subject food to decomposition, that indicate that food is contaminated with filth, or that otherwise may cause food to be adulterated.

Mixed-type facility means an establishment that engages in both activities that are exempt from registration under section 415 of the Federal Food, Drug, and Cosmetic Act and activities that require the establishment to be registered. An example of such a facility is a "farm mixed-type facility," which is an establishment that is a farm, but also conducts activities outside the farm definition that require the establishment to be registered.

Monitor means to conduct a planned sequence of observations or measurements to assess whether control measures are operating as intended.

Packing means placing food into a container other than packaging the food and also includes re-packing and activities performed incidental to packing or re-packing a food (*e.g.,* activities performed for the safe or effective packing or re-packing of that food (such as sorting, culling, grading, and weighing or conveying incidental to packing or re-packing), but does not include activities that transform a raw agricultural commodity into a processed food as defined in section 201(gg) of the Federal Food, Drug, and Cosmetic Act.

Pathogen means a microorganism of public health significance.

Pest refers to any objectionable animals or insects including birds, rodents, flies, and larvae.

Plant means the building or structure or parts thereof, used for or in connection with the manufacturing, processing, packing, or holding of human food.

Preventive controls means those riskbased, reasonably appropriate procedures, practices, and processes that a person knowledgeable about the safe manufacturing, processing, packing, or holding of food would employ to significantly minimize or prevent the hazards identified under the hazard analysis that are consistent with the current scientific understanding of safe food manufacturing, processing, packing, or holding at the time of the analysis.

Preventive controls qualified individual means a qualified individual who has successfully completed training in the development and application of risk-based preventive controls at least equivalent to that received under a standardized curriculum recognized as adequate by FDA or is otherwise qualified through job experience to develop and apply a food safety system.

Qualified auditor means a person who is a qualified individual as defined in this part and has technical expertise obtained through education, training, or experience (or a combination thereof) necessary to perform the auditing function as required by § 117.180(c)(2). Examples of potential qualified auditors include:

1) A government employee, including a foreign government employee; and
2) An audit agent of a certification body that is accredited in accordance with regulations in part 1, subpart M of this chapter.

Qualified end-user, with respect to a food, means the consumer of the food (where the term consumer does not include a business); or a restaurant or retail food establishment (as those terms are defined in § 1.227 of this chapter) that:

1) Is located;
   i) In the same State or the same Indian reservation as the qualified facility that sold the food to such restaurant or establishment; or
   ii) Not more than 275 miles from such facility; and
2) Is purchasing the food for sale directly to consumers at such restaurant or retail food establishment.

Qualified facility means (when including the sales by any subsidiary; affiliate; or subsidiaries or affiliates, collectively, of any entity of which the facility is a subsidiary or affiliate) a facility that is a very small business as defined in this part, or a facility to which both of the following apply:

1) During the 3-year period preceding the applicable calendar year, the average annual monetary value of the food manufactured, processed, packed or held at such facility that is sold directly to qualified end-users (as defined in this part) during such period exceeded the average annual monetary value of the food sold by such facility to all other purchasers; and
2) The average annual monetary value of all food sold during the 3-year period preceding the applicable calendar year was less than $500,000, adjusted for inflation.

Qualified facility exemption means an exemption applicable to a qualified facility under § 117.5(a).

Qualified individual means a person who has the education, training, or experience (or a combination thereof) necessary to manufacture, process, pack, or hold clean and safe food as appropriate to the individual's assigned duties. A qualified individual may be, but is not required to be, an employee of the establishment.

Quality control operation means a planned and systematic procedure for taking all actions necessary to prevent food from being adulterated.

Raw agricultural commodity has the meaning given in section 201(r) of the Federal Food, Drug, and Cosmetic Act.

Ready-to-eat food (RTE food) means any food that is normally eaten in its raw state or any other food, including a processed food, for which it is reasonably foreseeable that the food will be eaten without further processing that would significantly minimize biological hazards.

Receiving facility means a facility that is subject to subparts C and G of this part and that manufactures/processes a raw material or other ingredient that it receives from a supplier.

Rework means clean, unadulterated food that has been removed from processing for reasons other than insanitary conditions or that has been successfully reconditioned by reprocessing and that is suitable for use as food.

Safe-moisture level is a level of moisture low enough to prevent the growth of undesirable microorganisms in the finished product under the intended conditions of manufacturing, processing, packing, and holding. The safe moisture level for a food is related to its water activity (aw). An aw will be considered safe for a food if adequate data are available that demonstrate that the food at or below the given aw will not support the growth of undesirable microorganisms.

Sanitize means to adequately treat cleaned surfaces by a process that is effective in destroying vegetative cells of pathogens, and in substantially reducing numbers of other undesirable microorganisms, but without adversely affecting the product or its safety for the consumer.

Significantly minimize means to reduce to an acceptable level, including to eliminate.

Small business means, for purposes of this part, a business employing fewer than 500 full-time equivalent employees.

Subsidiary means any company which is owned or controlled directly or indirectly by another company.

Supplier means the establishment that manufactures/processes the food, raises the animal, or grows the food that is provided to a receiving facility without further manufacturing/processing by another establishment, except for further manufacturing/processing that consists solely of the addition of labeling or similar activity of a *de minimis* nature.

Supply-chain-applied control means a preventive control for a hazard in a raw material or other ingredient when the hazard in the raw material or other ingredient is controlled before its receipt.

Unexposed packaged food means packaged food that is not exposed to the environment.

Validation means obtaining and evaluating scientific and technical evidence that a control measure, combination of control measures, or the food safety plan as a whole, when properly implemented, is capable of effectively controlling the identified hazards.

Verification means the application of methods, procedures, tests and other evaluations, in addition to monitoring, to determine whether a control measure or combination of control measures is or has been operating as intended and to establish the validity of the food safety plan.

Very small business means, for purposes of this part, a business (including any subsidiaries and affiliates) averaging less than $1,000,000, adjusted for inflation, per year, during the 3-year period preceding the applicable calendar year in sales of human food plus the market value of human food manufactured, processed, packed, or held without sale (*e.g.*, held for a fee).

Water activity (aw) is a measure of the free moisture in a food and is the quotient of the water vapor pressure of the substance divided by the vapor pressure of pure water at the same temperature.

Written procedures for receiving raw materials and other ingredients means written procedures to ensure that raw materials and other ingredients are received only from suppliers approved by the receiving facility (or, when necessary and appropriate, on a temporary basis from unapproved suppliers whose raw materials or other ingredients are subjected to adequate verification activities before acceptance for use).

You means, for purposes of this part, the owner, operator, or agent in charge of a facility.

§ 117.4 Qualifications of individuals who manufacture, process, pack, or hold food.

a) *Applicability.*
   1) The management of an establishment must ensure that all individuals who manufacture, process, pack, or hold food subject to subparts B and F of this part are qualified to perform their assigned duties.
   2) The owner, operator, or agent in charge of a facility must ensure that all individuals who manufacture, process, pack, or hold food subject to subpart C, D, E, F, or G of this part are qualified to perform their assigned duties.

b) *Qualifications of all individuals engaged in manufacturing, processing, packing, or holding food.* Each individual engaged in manufacturing, processing, packing, or holding food (including temporary and seasonal personnel) or in the supervision thereof must:
   1) Be a qualified individual as that term is defined in § 117.3—*i.e.*, have the education, training, or experience (or a combination thereof) necessary to manufacture, process, pack, or hold clean and safe food as appropriate to the individual's assigned duties; and
   2) Receive training in the principles of food hygiene and food safety, including the importance of employee health and personal hygiene, as appropriate to the food, the facility and the individual's assigned duties.

c) *Additional qualifications of supervisory personnel.* Responsibility for ensuring compliance by individuals with the requirements of this part must be clearly assigned to supervisory personnel who have the education, training, or experience (or a combination thereof) necessary to supervise the production of clean and safe food.

d) *Records.* Records that document training required by paragraph (b)(2) of this section must be established and maintained.

§ 117.5 Exemptions.

a) Except as provided by subpart E of this part, subparts C and G of this part does not apply to a qualified facility. Qualified facilities are subject to the modified requirements in § 117.201.

b) Subparts C and G of this part do not apply with respect to activities that are subject to part 123 of this chapter (Fish and Fishery Products) at a facility if you are required to comply with, and are in compliance with, part 123 of this chapter with respect to such activities.

c) Subparts C and G of this part do not apply with respect to activities that are subject to part 120 of this chapter (Hazard Analysis and Critical Control Point

(HACCP) Systems) at a facility if you are required to comply with, and are in compliance with, part 120 of this chapter with respect to such activities.

d) 1) Subparts C and G of this part do not apply with respect to activities that are subject to part 113 of this chapter (Thermally Processed Low-Acid Foods Packaged in Hermetically Sealed Containers) at a facility if you are required to comply with, and are in compliance with, part 113 of this chapter with respect to such activities.
   2) The exemption in paragraph (d)(1) of this section is applicable only with respect to the microbiological hazards that are regulated under part 113 of this chapter.

e) Subparts C and G do not apply to any facility with regard to the manufacturing, processing, packaging, or holding of a dietary supplement that is in compliance with the requirements of part 111 of this chapter (Current Good Manufacturing Practice in Manufacturing, Packaging, Labeling, or Holding Operations for Dietary Supplements) and section 761 of the Federal Food, Drug, and Cosmetic Act (Serious Adverse Event Reporting for Dietary Supplements).

f) Subparts C and G of this part do not apply to activities of a facility that are subject to section 419 of the Federal Food, Drug, and Cosmetic Act (Standards for Produce Safety).

g) 1) The exemption in paragraph (g)(3) of this section applies to packing or holding of processed foods on a farm mixed-type facility, except for processed foods produced by drying/dehydrating raw agricultural commodities to create a distinct commodity (such as drying/dehydrating grapes to produce raisins, and drying/dehydrating fresh herbs to produce dried herbs), and packaging and labeling such commodities, without additional manufacturing/processing (such as chopping and slicing), the packing and holding of which are within the "farm" definition in § 1.227 of this chapter. Activities that are within the "farm" definition, when conducted on a farm mixed-type facility, are not subject to the requirements of subparts C and G of this part and therefore do not need to be specified in the exemption.
   2) For the purposes of paragraphs (g)(3) and (h)(3) of this section, the following terms describe the foods associated with the activity/food combinations. Several foods that are fruits or vegetables are separately considered for the purposes of these activity/food combinations (*i.e.*, coffee beans, cocoa beans, fresh herbs, peanuts, sugarcane, sugar beets, tree nuts, seeds for direct consumption) to appropriately address specific hazards associated with these foods and/or processing activities conducted on these foods.
      i) *Dried/dehydrated fruit and vegetable products* includes only those processed food products such as raisins and dried legumes made

without additional manufacturing/processing beyond drying/dehydrating, packaging, and/or labeling.

ii) *Other fruit and vegetable products* includes those processed food products that have undergone one or more of the following processes: acidification, boiling, canning, coating with things other than wax/oil/resin, cooking, cutting, chopping, grinding, peeling, shredding, slicing, or trimming. Examples include flours made from legumes (such as chickpea flour), pickles, and snack chips made from potatoes or plantains. Examples also include dried fruit and vegetable products made with additional manufacturing/processing (such as dried apple slices; pitted, dried plums, cherries, and apricots; and sulfited raisins). This category does not include dried/dehydrated fruit and vegetable products made without additional manufacturing/processing as described in paragraph (g)(2)(i) of this section. This category also does not include products that require time/temperature control for safety (such as fresh-cut fruits and vegetables).

iii) *Peanut and tree nut products* includes processed food products such as roasted peanuts and tree nuts, seasoned peanuts and tree nuts, and peanut and tree nut flours.

iv) *Processed seeds for direct consumption* include processed food products such as roasted pumpkin seeds, roasted sunflower seeds, and roasted flax seeds.

v) *Dried/dehydrated herb and spice products* includes only processed food products such as dried intact herbs made without additional manufacturing/processing beyond drying/dehydrating, packaging, and/or labeling.

vi) *Other herb and spice products* includes those processed food products such as chopped fresh herbs, chopped or ground dried herbs (including tea), herbal extracts (*e.g.,* essential oils, extracts containing more than 20 percent ethanol, extracts containing more than 35 percent glycerin), dried herb- or spice-infused honey, and dried herb- or spice-infused oils and/or vinegars. This category does not include dried/dehydrated herb and spice products made without additional manufacturing/processing beyond drying/dehydrating, packaging, and/or labeling as described in paragraph (g)(2)(v) of this section. This category also does not include products that require time/temperature control for safety, such as fresh herb-infused oils.

vii) *Grains* include barley, dent- or flint-corn, sorghum, oats, rice, rye, wheat, amaranth, quinoa, buckwheat and oilseeds for oil extraction (such as cotton seed, flax seed, rapeseed, soybeans, and sunflower seed).

viii) *Milled grain products* include processed food products such as flour, bran, and corn meal.

ix) *Baked goods* include processed food products such as breads, brownies, cakes, cookies, and crackers. This category does not include products that require time/temperature control for safety, such as cream-filled pastries.

x) *Other grain products* include processed food products such as dried cereal, dried pasta, oat flakes, and popcorn. This category does not include milled grain products as described in paragraph (g)(2)(viii) of this section or baked goods as described in paragraph (g)(2)(ix) of this section.

3) Subparts C and G of this part do not apply to on-farm packing or holding of food by a small or very small business, and § 117.201 does not apply to on-farm packing or holding of food by a very small business, if the only packing and holding activities subject to section 418 of the Federal Food, Drug, and Cosmetic Act that the business conducts are the following low-risk packing or holding activity/food combinations—*i.e.*, packing (or repacking) (including weighing or conveying incidental to packing or repacking); sorting, culling, or grading incidental to packing or storing; and storing (ambient, cold and controlled atmosphere) of:

i) Baked goods (*e.g.*, bread and cookies);
ii) Candy (*e.g.*, hard candy, fudge, maple candy, maple cream, nut brittles, taffy, and toffee);
iii) Cocoa beans (roasted);
iv) Cocoa products;
v) Coffee beans (roasted);
vi) Game meat jerky;
vii) Gums, latexes, and resins that are processed foods;
viii) Honey (pasteurized);
ix) Jams, jellies, and preserves;
x) Milled grain products (*e.g.*, flour, bran, and corn meal);
xi) Molasses and treacle;
xii) Oils (*e.g.*, olive oil and sunflower seed oil);
xiii) Other fruit and vegetable products (*e.g.*, flours made from legumes; pitted, dried fruits; sliced, dried apples; snack chips);
xiv) Other grain products (*e.g.*, dried pasta, oat flakes, and popcorn);
xv) Other herb and spice products (*e.g.*, chopped or ground dried herbs, herbal extracts);
xvi) Peanut and tree nut products (*e.g.*, roasted peanuts and tree nut flours);
xvii) Processed seeds for direct consumption (*e.g.*, roasted pumpkin seeds);
xviii) Soft drinks and carbonated water;
xix) Sugar;
xx) Syrups (*e.g.*, maple syrup and agave syrup);

xxi) Trail mix and granola;

xxii) Vinegar; and

xxiii) Any other processed food that does not require time/temperature control for safety (e..g., vitamins, minerals, and dietary ingredients (*e.g.,* bone meal) in powdered, granular, or other solid form).

h) 1) The exemption in paragraph (h)(3) of this section applies to manufacturing/processing of foods on a farm mixed-type facility, except for manufacturing/processing that is within the "farm" definition in § 1.227 of this chapter. Drying/dehydrating raw agricultural commodities to create a distinct commodity (such as drying/ dehydrating grapes to produce raisins, and drying/dehydrating fresh herbs to produce dried herbs), and packaging and labeling such commodities, without additional manufacturing/processing (such as chopping and slicing), are within the "farm" definition in § 1.227 of this chapter. In addition, treatment to manipulate ripening of raw agricultural commodities (such as by treating produce with ethylene gas), and packaging and labeling the treated raw agricultural commodities, without additional manufacturing/processing, is within the "farm" definition. In addition, coating intact fruits and vegetables with wax, oil, or resin used for the purpose of storage or transportation is within the "farm" definition. Activities that are within the "farm" definition, when conducted on a farm mixed-type facility, are not subject to the requirements of subparts C and G of this part and therefore do not need to be specified in the exemption.

2) The terms in paragraph (g)(2) of this section describe certain foods associated with the activity/food combinations in paragraph (h)(3) of this section.

3) Subparts C and G of this part do not apply to on-farm manufacturing/ processing activities conducted by a small or very small business for distribution into commerce, and § 117.201 does not apply to on-farm manufacturing/processing activities conducted by a very small business for distribution into commerce, if the only manufacturing/processing activities subject to section 418 of the Federal Food, Drug, and Cosmetic Act that the business conducts are the following low-risk manufacturing/processing activity/food combinations:

i) Boiling gums, latexes, and resins;

ii) Chopping, coring, cutting, peeling, pitting, shredding, and slicing acid fruits and vegetables that have a pH less than 4.2 (*e.g.,* cutting lemons and limes), baked goods (*e.g.,* slicing bread), dried/ dehydrated fruit and vegetable products (*e.g.,* pitting dried plums), dried herbs and other spices (*e.g.,* chopping intact, dried basil), game meat jerky, gums/ latexes/resins, other grain products (*e.g.,* shredding dried cereal), peanuts and tree nuts, and peanut and tree nut products (*e.g.,* chopping roasted peanuts);

iii) Coating dried/dehydrated fruit and vegetable products (*e.g.,* coating raisins with chocolate), other fruit and vegetable products except for non-dried, non-intact fruits and vegetables (*e.g.,* coating dried plum pieces, dried pitted cherries, and dried pitted apricots with chocolate are low-risk activity/food combinations but coating apples on a stick with caramel is not a low-risk activity/food combination), other grain products (*e.g.,* adding caramel to popcorn or adding seasonings to popcorn provided that the seasonings have been treated to significantly minimize pathogens, peanuts and tree nuts (*e.g.,* adding seasonings provided that the seasonings have been treated to significantly minimize pathogens), and peanut and tree nut products (*e.g.,* adding seasonings provided that the seasonings have been treated to significantly minimize pathogens));

iv) Drying/dehydrating (that includes additional manufacturing or is performed on processed foods) other fruit and vegetable products with pH less than 4.2 (*e.g.,* drying cut fruit and vegetables with pH less than 4.2), and other herb and spice products (*e.g.,* drying chopped fresh herbs, including tea);

v) Extracting (including by pressing, by distilling, and by solvent extraction) from dried/dehydrated herb and spice products (*e.g.,* dried mint), fresh herbs (*e.g.,* fresh mint), fruits and vegetables (*e.g.,* olives, avocados), grains (*e.g.,* oilseeds), and other herb and spice products (*e.g.,* chopped fresh mint, chopped dried mint);

vi) Freezing acid fruits and vegetables with pH less than 4.2 and other fruit and vegetable products with pH less than 4.2 (*e.g.,* cut fruits and vegetables);

vii) Grinding/cracking/crushing/ milling baked goods (*e.g.,* crackers), cocoa beans (roasted), coffee beans (roasted), dried/dehydrated fruit and vegetable products (*e.g.,* raisins and dried legumes), dried/ dehydrated herb and spice products (*e.g.,* intact dried basil), grains (*e.g.,* oats, rice, rye, wheat), other fruit and vegetable products (*e.g.,* dried, pitted dates), other grain products (*e.g.,* dried cereal), other herb and spice products (*e.g.,* chopped dried herbs), peanuts and tree nuts, and peanut and tree nut products (*e.g.,* roasted peanuts);

viii) Labeling baked goods that do not contain food allergens, candy that does not contain food allergens, cocoa beans (roasted), cocoa products that do not contain food allergens), coffee beans (roasted), game meat jerky, gums/ latexes/resins that are processed foods, honey (pasteurized), jams/jellies/ preserves, milled grain products that do not contain food allergens (*e.g.,* corn meal) or that are single-ingredient foods (*e.g.,* wheat flour, wheat bran), molasses and treacle, oils, other fruit and vegetable products that do not contain food allergens (*e.g.,* snack chips made from potatoes or plantains), other grain products that do not contain food allergens (*e.g.*, popcorn), other herb and spice products (*e.g.*, chopped or ground dried herbs), peanut or tree nut products, (provided that they are single ingredient, or are in forms in which the consumer can reasonably be expected to recognize the food allergen(s) without label declaration, or both (*e.g.*, roasted or seasoned whole nuts, single ingredient peanut or tree nut flours)), processed seeds for direct consumption, soft drinks and carbonated water, sugar, syrups, trail mix and granola (other than those containing milk chocolate and provided that peanuts and/or tree nuts are in forms in which the consumer can reasonably be expected to recognize the food allergen(s) without label declaration), vinegar, and any other processed food that does not require time/temperature control for safety and that does not contain food allergens (*e.g.*, vitamins, minerals, and dietary ingredients (*e.g.*, bone meal) in powdered, granular, or other solid form);

ix) Making baked goods from milled grain products (*e.g.*, breads and cookies);

x) Making candy from peanuts and tree nuts (*e.g.*, nut brittles), sugar/syrups (*e.g.*, taffy, toffee), and saps (*e.g.*, maple candy, maple cream);

xi) Making cocoa products from roasted cocoa beans;

xii) Making dried pasta from grains;

xiii) Making jams, jellies, and preserves from acid fruits and vegetables with a pH of 4.6 or below;

xiv) Making molasses and treacle from sugar beets and sugarcane;

xv) Making oat flakes from grains;

xvi) Making popcorn from grains;

xvii) Making snack chips from fruits and vegetables (*e.g.*, making plantain and potato chips);

xviii) Making soft drinks and carbonated water from sugar, syrups, and water;

xix) Making sugars and syrups from fruits and vegetables (*e.g.*, dates), grains (*e.g.*, rice, sorghum), other grain products (*e.g.*, malted grains such as barley), saps (*e.g.*, agave, birch, maple, palm), sugar beets, and sugarcane;

xx) Making trail mix and granola from cocoa products (*e.g.*, chocolate), dried/dehydrated fruit and vegetable products (*e.g.*, raisins), other fruit and vegetable products (*e.g.*, chopped dried fruits), other grain products (*e.g.*, oat flakes), peanut and tree nut products, and processed seeds for direct consumption, provided that peanuts, tree nuts, and processed seeds are treated to significantly minimize pathogens;

xxi) Making vinegar from fruits and vegetables, other fruit and vegetable products (*e.g.*, fruit wines, apple cider), and other grain products (*e.g.*, malt);

xxii) Mixing baked goods (*e.g.*, types of cookies), candy (*e.g.*, varieties of taffy), cocoa beans (roasted), coffee beans (roasted), dried/dehydrated fruit and vegetable products (*e.g.*, dried blueberries, dried currants, and raisins), dried/dehydrated herb and spice products (*e.g.*, dried, intact basil and dried, intact oregano), honey (pasteurized), milled grain products (*e.g.*, flour, bran, and corn meal), other fruit and vegetable products (*e.g.*, dried, sliced apples and dried, sliced peaches), other grain products (*e.g.*, different types of dried pasta), other herb and spice products (*e.g.*, chopped or ground dried herbs, dried herb- or spice-infused honey, and dried herb- or spice-infused oils and/or vinegars), peanut and tree nut products, sugar, syrups, vinegar, and any other processed food that does not require time/temperature control for safety (*e.g.*, vitamins, minerals, and dietary ingredients (*e.g.*, bone meal) in powdered, granular, or other solid form);

xxiii) Packaging baked goods (*e.g.*, bread and cookies), candy, cocoa beans (roasted), cocoa products, coffee beans (roasted), game meat jerky, gums/ latexes/resins that are processed foods, honey (pasteurized), jams/jellies/ preserves, milled grain products (*e.g.*, flour, bran, corn meal), molasses and treacle, oils, other fruit and vegetable products (*e.g.*, pitted, dried fruits; sliced, dried apples; snack chips), other grain products (*e.g.*, popcorn), other herb and spice products (*e.g.*, chopped or ground dried herbs), peanut and tree nut products, processed seeds for direct consumption, soft drinks and carbonated water, sugar, syrups, trail mix and granola, vinegar, and any other processed food that does not require time/temperature control for safety (*e.g.*, vitamins, minerals, and dietary ingredients (*e.g.*, bone meal) in powdered, granular, or other solid form);

xxiv) Pasteurizing honey;

xxv) Roasting and toasting baked goods (*e.g.*, toasting bread for croutons);

xxvi) Salting other grain products (*e.g.*, soy nuts), peanut and tree nut products, and processed seeds for direct consumption; and

xxvii) Sifting milled grain products (*e.g.*, flour, bran, corn meal), other fruit and vegetable products (*e.g.*, chickpea flour), and peanut and tree nut products (*e.g.*, peanut flour, almond flour).

i) 1) Subparts C and G of this part do not apply with respect to alcoholic beverages at a facility that meets the following two conditions:

i) Under the Federal Alcohol Administration Act (27 U.S.C. 201 *et seq.*) or chapter 51 of subtitle E of the Internal Revenue Code of 1986 (26 U.S.C. 5001 *et seq.*) the facility is required to obtain a permit from, register with, or obtain approval of a notice or application from the Secretary of the Treasury as a condition of doing business in the

United States, or is a foreign facility of a type that would require such a permit, registration, or approval if it were a domestic facility; and

ii) Under section 415 of the Federal Food, Drug, and Cosmetic Act the facility is required to register as a facility because it is engaged in manufacturing, processing, packing, or holding one or more alcoholic beverages.

2) Subparts C and G of this part do not apply with respect to food that is not an alcoholic beverage at a facility described in paragraph (i)(1) of this section, provided such food:

i) Is in prepackaged form that prevents any direct human contact with such food; and

ii) Constitutes not more than 5 percent of the overall sales of the facility, as determined by the Secretary of the Treasury.

j) Subparts C and G of this part do not apply to facilities that are solely engaged in the storage of raw agricultural commodities (other than fruits and vegetables) intended for further distribution or processing.

k) 1) Except as provided by paragraph (k)(2) of this section, subpart B of this part does not apply to any of the following:

i) "Farms" (as defined in § 1.227 of this chapter);

ii) Fishing vessels that are not subject to the registration requirements of part 1, subpart H of this chapter in accordance with § 1.226(f) of this chapter;

iii) Establishments solely engaged in the holding and/or transportation of one or more raw agricultural commodities;

iv) Activities of "farm mixed-type facilities" (as defined in § 1.227 of this chapter) that fall within the definition of "farm"; or

v) Establishments solely engaged in hulling, shelling, drying, packing, and/ or holding nuts (without additional manufacturing/processing, such as roasting nuts).

2) If a "farm" or "farm mixed-type facility" dries/dehydrates raw agricultural commodities that are produce as defined in part 112 of this chapter to create a distinct commodity, subpart B of this part applies to the packaging, packing, and holding of the dried commodities. Compliance with this requirement may be achieved by complying with subpart B of this part or with the applicable requirements for packing and holding in part 112 of this chapter.

§ 117.7 Applicability of subparts C, D, and G of this part to a facility solely engaged in the storage of unexposed packaged food.

a) *Applicability of subparts C and G.* Subparts C and G of this part do not apply to a facility solely engaged in the storage of unexposed packaged food.

b) *Applicability of subpart D.* A facility solely engaged in the storage of unexposed packaged food, including unexposed packaged food that requires time/temperature control to significantly minimize or prevent the growth of, or toxin production by, pathogens is subject to the modified requirements in § 117.206 for any unexposed packaged food that requires time/temperature control to significantly minimize or prevent the growth of, or toxin production by, pathogens.

§ 117.8 Applicability of subpart B of this part to the off-farm packing and holding of raw agricultural commodities.
Subpart B of this part applies to the off-farm packaging, packing, and holding of raw agricultural commodities. Compliance with this requirement for raw agricultural commodities that are produce as defined in part 112 of this chapter may be achieved by complying with subpart B of this part or with the applicable requirements for packing and holding in part 112 of this chapter.

§ 117.9 Records required for this subpart.

a) Records that document training required by § 117.4(b)(2) must be established and maintained.
b) The records that must be established and maintained are subject to the requirements of subpart F of this part.

Subpart B—Current Good Manufacturing Practice

§ 117.10 Personnel.
The management of the establishment must take reasonable measures and precautions to ensure the following:

a) *Disease control.* Any person who, by medical examination or supervisory observation, is shown to have, or appears to have, an illness, open lesion, including boils, sores, or infected wounds, or any other abnormal source of microbial contamination by which there is a reasonable possibility of food, food-contact surfaces, or food-packaging materials becoming contaminated, must be excluded from any operations which may be expected to result in such contamination until the condition is corrected, unless conditions such as open lesions, boils, and infected wounds are adequately covered (*e.g.*, by an impermeable cover). Personnel must be instructed to report such health conditions to their supervisors.
b) *Cleanliness.* All persons working in direct contact with food, food-contact surfaces, and food-packaging materials must conform to hygienic practices while on duty to the extent necessary to protect against allergen cross-contact and against contamination of food. The methods for maintaining cleanliness include:
    1) Wearing outer garments suitable to the operation in a manner that protects against allergen cross-contact and against the contamination of food, foodcontact surfaces, or food-packaging materials.
    2) Maintaining adequate personal cleanliness.

3) Washing hands thoroughly (and sanitizing if necessary to protect against contamination with undesirable microorganisms) in an adequate hand-washing facility before starting work, after each absence from the work station, and at any other time when the hands may have become soiled or contaminated.
4) Removing all unsecured jewelry and other objects that might fall into food, equipment, or containers, and removing hand jewelry that cannot be adequately sanitized during periods in which food is manipulated by hand. If such hand jewelry cannot be removed, it may be covered by material which can be maintained in an intact, clean, and sanitary condition and which effectively protects against the contamination by these objects of the food, food-contact surfaces, or food-packaging materials.
5) Maintaining gloves, if they are used in food handling, in an intact, clean, and sanitary condition.
6) Wearing, where appropriate, in an effective manner, hair nets, headbands, caps, beard covers, or other effective hair restraints.
7) Storing clothing or other personal belongings in areas other than where food is exposed or where equipment or utensils are washed.
8) Confining the following to areas other than where food may be exposed or where equipment or utensils are washed: eating food, chewing gum, drinking beverages, or using tobacco. (9) Taking any other necessary precautions to protect against allergen cross-contact and against contamination of food, food-contact surfaces, or food packaging materials with microorganisms or foreign substances (including perspiration, hair, cosmetics, tobacco, chemicals, and medicines applied to the skin).

§ 117.20 Plant and grounds.

a) *Grounds.* The grounds about a food plant under the control of the operator must be kept in a condition that will protect against the contamination of food. The methods for adequate maintenance of grounds must include:
   1) Properly storing equipment, removing litter and waste, and cutting weeds or grass within the immediate vicinity of the plant that may constitute an attractant, breeding place, or harborage for pests.
   2) Maintaining roads, yards, and parking lots so that they do not constitute a source of contamination in areas where food is exposed.
   3) Adequately draining areas that may contribute contamination to food by seepage, foot-borne filth, or providing a breeding place for pests.
   4) Operating systems for waste treatment and disposal in an adequate manner so that they do not constitute a source of contamination in areas where food is exposed.
   5) If the plant grounds are bordered by grounds not under the operator's control and not maintained in the manner described in paragraphs (a)(1)

through (4) of this section, care must be exercised in the plant by inspection, extermination, or other means to exclude pests, dirt, and filth that may be a source of food contamination.

b) *Plant construction and design.* The plant must be suitable in size, construction, and design to facilitate maintenance and sanitary operations for food-production purposes (*i.e.*, manufacturing, processing, packing, and holding). The plant must:
   1) Provide adequate space for such placement of equipment and storage of materials as is necessary for maintenance, sanitary operations, and the production of safe food.
   2) Permit the taking of adequate precautions to reduce the potential for allergen cross-contact and for contamination of food, food-contact surfaces, or food-packaging materials with microorganisms, chemicals, filth, and other extraneous material. The potential for allergen cross-contact and for contamination may be reduced by adequate food safety controls and operating practices or effective design, including the separation of operations in which allergen cross-contact and contamination are likely to occur, by one or more of the following means: location, time, partition, air flow systems, dust control systems, enclosed systems, or other effective means.
   3) Permit the taking of adequate precautions to protect food in installed outdoor bulk vessels by any effective means, including:
      i) Using protective coverings.
      ii) Controlling areas over and around the vessels to eliminate harborages for pests.
      iii) Checking on a regular basis for pests and pest infestation.
      iv) Skimming fermentation vessels, as necessary.
   4) Be constructed in such a manner that floors, walls, and ceilings may be adequately cleaned and kept clean and kept in good repair; that drip or condensate from fixtures, ducts and pipes does not contaminate food, foodcontact surfaces, or food-packaging materials; and that aisles or working spaces are provided between equipment and walls and are adequately unobstructed and of adequate width to permit employees to perform their duties and to protect against contaminating food, food-contact surfaces, or food-packaging materials with clothing or personal contact.
   5) Provide adequate lighting in handwashing areas, dressing and locker rooms, and toilet rooms and in all areas where food is examined, manufactured, processed, packed, or held and where equipment or utensils are cleaned; and provide shatter-resistant light bulbs, fixtures, skylights, or other glass suspended over exposed food in any step of preparation or otherwise protect against food contamination in case of glass breakage.
   6) Provide adequate ventilation or control equipment to minimize dust, odors and vapors (including steam and noxious fumes) in areas where

they may cause allergen cross-contact or contaminate food; and locate and operate fans and other air-blowing equipment in a manner that minimizes the potential for allergen cross-contact and for contaminating food, food-packaging materials, and food-contact surfaces.

7) Provide, where necessary, adequate screening or other protection against pests.

§ 117.35 Sanitary operations.

a) *General maintenance.* Buildings, fixtures, and other physical facilities of the plant must be maintained in a clean and sanitary condition and must be kept in repair adequate to prevent food from becoming adulterated. Cleaning and sanitizing of utensils and equipment must be conducted in a manner that protects against allergen cross-contact and against contamination of food, foodcontact surfaces, or food-packaging materials.

b) *Substances used in cleaning and sanitizing; storage of toxic materials.*
   1) Cleaning compounds and sanitizing agents used in cleaning and sanitizing procedures must be free from undesirable microorganisms and must be safe and adequate under the conditions of use. Compliance with this requirement must be verified by any effective means, including purchase of these substances under a letter of guarantee or certification or examination of these substances for contamination. Only the following toxic materials may be used or stored in a plant where food is processed or exposed:
      i) Those required to maintain clean and sanitary conditions;
      ii) Those necessary for use in laboratory testing procedures;
      iii) Those necessary for plant and equipment maintenance and operation; and
      iv) Those necessary for use in the plant's operations.
   2) Toxic cleaning compounds, sanitizing agents, and pesticide chemicals must be identified, held, and stored in a manner that protects against contamination of food, food-contact surfaces, or food-packaging materials.

c) *Pest control.* Pests must not be allowed in any area of a food plant. Guard, guide, or pest-detecting dogs may be allowed in some areas of a plant if the presence of the dogs is unlikely to result in contamination of food, foodcontact surfaces, or food-packaging materials. Effective measures must be taken to exclude pests from the manufacturing, processing, packing, and holding areas and to protect against the contamination of food on the premises by pests. The use of pesticides to control pests in the plant is permitted only under precautions and restrictions that will protect against the contamination of food, food-contact surfaces, and food- packaging materials.

d) *Sanitation of food-contact surfaces.* All food-contact surfaces, including utensils and food-contact surfaces of equipment, must be cleaned as frequently as necessary to protect against allergen cross-contact and against contamination of food.

1) Food-contact surfaces used for manufacturing/processing, packing, or holding low-moisture food must be in a clean, dry, sanitary condition before use. When the surfaces are wet-cleaned, they must, when necessary, be sanitized and thoroughly dried before subsequent use.
2) In wet processing, when cleaning is necessary to protect against allergen cross-contact or the introduction of microorganisms into food, all food-contact surfaces must be cleaned and sanitized before use and after any interruption during which the foodcontact surfaces may have become contaminated. Where equipment and utensils are used in a continuous production operation, the utensils and food-contact surfaces of the equipment must be cleaned and sanitized as necessary.
3) Single-service articles (such as utensils intended for one-time use, paper cups, and paper towels) must be stored, handled, and disposed of in a manner that protects against allergen cross-contact and against contamination of food, food-contact surfaces, or food-packaging materials.

e) *Sanitation of non-food-contact surfaces.* Non-food-contact surfaces of equipment used in the operation of a food plant must be cleaned in a manner and as frequently as necessary to protect against allergen cross-contact and against contamination of food, foodcontact surfaces, and food-packaging materials.
f) *Storage and handling of cleaned portable equipment and utensils.* Cleaned and sanitized portable equipment with food-contact surfaces and utensils must be stored in a location and manner that protects food-contact surfaces from allergen cross-contact and from contamination.

§ 117.37 Sanitary facilities and controls.

Each plant must be equipped with adequate sanitary facilities and accommodations including:

a) *Water supply.* The water supply must be adequate for the operations intended and must be derived from an adequate source. Any water that contacts food, food-contact surfaces, or food-packaging materials must be safe and of adequate sanitary quality. Running water at a suitable temperature, and under pressure as needed, must be provided in all areas where required for the processing of food, for the cleaning of equipment, utensils, and food-packaging materials, or for employee sanitary facilities.
b) *Plumbing.* Plumbing must be of adequate size and design and adequately installed and maintained to:
    1) Carry adequate quantities of water to required locations throughout the plant.
    2) Properly convey sewage and liquid disposable waste from the plant.

3) Avoid constituting a source of contamination to food, water supplies, equipment, or utensils or creating an unsanitary condition.
4) Provide adequate floor drainage in all areas where floors are subject to flooding-type cleaning or where normal operations release or discharge water or other liquid waste on the floor.
5) Provide that there is not backflow from, or cross-connection between, piping systems that discharge waste water or sewage and piping systems that carry water for food or food manufacturing.

c) *Sewage disposal.* Sewage must be disposed of into an adequate sewerage system or disposed of through other adequate means.
d) *Toilet facilities.* Each plant must provide employees with adequate, readily accessible toilet facilities. Toilet facilities must be kept clean and must not be a potential source of contamination of food, food-contact surfaces, or food-packaging materials.
e) *Hand-washing facilities.* Each plant must provide hand-washing facilities designed to ensure that an employee's hands are not a source of contamination of food, food-contact surfaces, or food-packaging materials, by providing facilities that are adequate, convenient, and furnish running water at a suitable temperature.
f) *Rubbish and offal disposal.* Rubbish and any offal must be so conveyed, stored, and disposed of as to minimize the development of odor, minimize the potential for the waste becoming an attractant and harborage or breeding place for pests, and protect against contamination of food, foodcontact surfaces, food-packaging materials, water supplies, and ground surfaces.

§ 117.40 Equipment and utensils.

a) 1) All plant equipment and utensils used in manufacturing, processing, packing, or holding food must be so designed and of such material and workmanship as to be adequately cleanable, and must be adequately maintained to protect against allergen cross-contact and contamination.
   2) Equipment and utensils must be designed, constructed, and used appropriately to avoid the adulteration of food with lubricants, fuel, metal fragments, contaminated water, or any other contaminants.
   3) Equipment must be installed so as to facilitate the cleaning and maintenance of the equipment and of adjacent spaces.
   4) Food-contact surfaces must be corrosion-resistant when in contact with food.
   5) Food-contact surfaces must be made of nontoxic materials and designed to withstand the environment of their intended use and the action of food, and, if applicable, cleaning compounds, sanitizing agents, and cleaning procedures.

6) Food-contact surfaces must be maintained to protect food from allergen cross-contact and from being contaminated by any source, including unlawful indirect food additives.

b) Seams on food-contact surfaces must be smoothly bonded or maintained so as to minimize accumulation of food particles, dirt, and organic matter and thus minimize the opportunity for growth of microorganisms and allergen cross-contact.
c) Equipment that is in areas where food is manufactured, processed, packed, or held and that does not come into contact with food must be so constructed that it can be kept in a clean and sanitary condition.
d) Holding, conveying, and manufacturing systems, including gravimetric, pneumatic, closed, and automated systems, must be of a design and construction that enables them to be maintained in an appropriate clean and sanitary condition.
e) Each freezer and cold storage compartment used to store and hold food capable of supporting growth of microorganisms must be fitted with an indicating thermometer, temperature measuring device, or temperature recording device so installed as to show the temperature accurately within the compartment.
f) Instruments and controls used for measuring, regulating, or recording temperatures, pH, acidity, water activity, or other conditions that control or prevent the growth of undesirable microorganisms in food must be accurate and precise and adequately maintained, and adequate in number for their designated uses.
g) Compressed air or other gases mechanically introduced into food or used to clean food-contact surfaces or equipment must be treated in such a way that food is not contaminated with unlawful indirect food additives.

§ 117.80 Processes and controls.

a) *General.*
   1) All operations in the manufacturing, processing, packing, and holding of food (including operations directed to receiving, inspecting, transporting, and segregating) must be conducted in accordance with adequate sanitation principles.
   2) Appropriate quality control operations must be employed to ensure that food is suitable for human consumption and that food-packaging materials are safe and suitable.
   3) Overall sanitation of the plant must be under the supervision of one or more competent individuals assigned responsibility for this function.
   4) Adequate precautions must be taken to ensure that production procedures do not contribute to allergen cross-contact and to contamination from any source.

5) Chemical, microbial, or extraneous-material testing procedures must be used where necessary to identify sanitation failures or possible allergen cross-contact and food contamination.
6) All food that has become contaminated to the extent that it is adulterated must be rejected, or if appropriate, treated or processed to eliminate the contamination.

b) *Raw materials and other ingredients.*
1) Raw materials and other ingredients must be inspected and segregated or otherwise handled as necessary to ascertain that they are clean and suitable for processing into food and must be stored under conditions that will protect against allergen cross-contact and against contamination and minimize deterioration. Raw materials must be washed or cleaned as necessary to remove soil or other contamination. Water used for washing, rinsing, or conveying food must be safe and of adequate sanitary quality. Water may be reused for washing, rinsing, or conveying food if it does not cause allergen cross-contact or increase the level of contamination of the food.
2) Raw materials and other ingredients must either not contain levels of microorganisms that may render the food injurious to the health of humans, or they must be pasteurized or otherwise treated during manufacturing operations so that they no longer contain levels that would cause the product to be adulterated.
3) Raw materials and other ingredients susceptible to contamination with aflatoxin or other natural toxins must comply with FDA regulations for poisonous or deleterious substances before these raw materials or other ingredients are incorporated into finished food.
4) Raw materials, other ingredients, and rework susceptible to contamination with pests, undesirable microorganisms, or extraneous material must comply with applicable FDA regulations for natural or unavoidable defects if a manufacturer wishes to use the materials in manufacturing food.
5) Raw materials, other ingredients, and rework must be held in bulk, or in containers designed and constructed so as to protect against allergen crosscontact and against contamination and must be held at such temperature and relative humidity and in such a manner as to prevent the food from becoming adulterated. Material scheduled for rework must be identified as such.
6) Frozen raw materials and other ingredients must be kept frozen. If thawing is required prior to use, it must be done in a manner that prevents the raw materials and other ingredients from becoming adulterated.
7) Liquid or dry raw materials and other ingredients received and stored in bulk form must be held in a manner that protects against allergen cross-contact and against contamination.

8) Raw materials and other ingredients that are food allergens, and rework that contains food allergens, must be identified and held in a manner that prevents allergen cross-contact.

c) *Manufacturing operations.*
   1) Equipment and utensils and food containers must be maintained in an adequate condition through appropriate cleaning and sanitizing, as necessary. Insofar as necessary, equipment must be taken apart for thorough cleaning.
   2) All food manufacturing, processing, packing, and holding must be conducted under such conditions and controls as are necessary to minimize the potential for the growth of microorganisms, allergen cross-contact, contamination of food, and deterioration of food.
   3) Food that can support the rapid growth of undesirable microorganisms must be held at temperatures that will prevent the food from becoming adulterated during manufacturing, processing, packing, and holding.
   4) Measures such as sterilizing, irradiating, pasteurizing, cooking, freezing, refrigerating, controlling pH, or controlling aw that are taken to destroy or prevent the growth of undesirable microorganisms must be adequate under the conditions of manufacture, handling, and distribution to prevent food from being adulterated.
   5) Work-in-process and rework must be handled in a manner that protects against allergen cross-contact, contamination, and growth of undesirable microorganisms.
   6) Effective measures must be taken to protect finished food from allergen cross-contact and from contamination by raw materials, other ingredients, or refuse. When raw materials, other ingredients, or refuse are unprotected, they must not be handled simultaneously in a receiving, loading, or shipping area if that handling could result in allergen cross-contact or contaminated food. Food transported by conveyor must be protected against allergen cross-contact and against contamination as necessary.
   7) Equipment, containers, and utensils used to convey, hold, or store raw materials and other ingredients, work-in-process, rework, or other food must be constructed, handled, and maintained during manufacturing, processing, packing, and holding in a manner that protects against allergen cross-contact and against contamination.
   8) Adequate measures must be taken to protect against the inclusion of metal or other extraneous material in food.
   9) Food, raw materials, and other ingredients that are adulterated:
      i) Must be disposed of in a manner that protects against the contamination of other food; or
      ii) If the adulterated food is capable of being reconditioned, it must be:
         A) Reconditioned (if appropriate) using a method that has been proven to be effective; or

B) Reconditioned (if appropriate) and reexamined and subsequently found not to be adulterated within the meaning of the Federal Food, Drug, and Cosmetic Act before being incorporated into other food.

10) Steps such as washing, peeling, trimming, cutting, sorting and inspecting, mashing, dewatering, cooling, shredding, extruding, drying, whipping, defatting, and forming must be performed so as to protect food against allergen cross-contact and against contamination. Food must be protected from contaminants that may drip, drain, or be drawn into the food.
11) Heat blanching, when required in the preparation of food capable of supporting microbial growth, must be effected by heating the food to the required temperature, holding it at this temperature for the required time, and then either rapidly cooling the food or passing it to subsequent manufacturing without delay. Growth and contamination by thermophilic microorganisms in blanchers must be minimized by the use of adequate operating temperatures and by periodic cleaning and sanitizing as necessary.
12) Batters, breading, sauces, gravies, dressings, dipping solutions, and other similar preparations that are held and used repeatedly over time must be treated or maintained in such a manner that they are protected against allergen cross-contact and against contamination, and minimizing the potential for the growth of undesirable microorganisms.
13) Filling, assembling, packaging, and other operations must be performed in such a way that the food is protected against allergen cross-contact, contamination and growth of undesirable microorganisms.
14) Food, such as dry mixes, nuts, intermediate moisture food, and dehydrated food, that relies principally on the control of aw for preventing the growth of undesirable microorganisms must be processed to and maintained at a safe moisture level.
15) Food, such as acid and acidified food, that relies principally on the control of pH for preventing the growth of undesirable microorganisms must be monitored and maintained at a pH of 4.6 or below.
16) When ice is used in contact with food, it must be made from water that is safe and of adequate sanitary quality in accordance with § 117.37(a), and must be used only if it has been manufactured in accordance with current good manufacturing practice as outlined in this part.

§ 117.93 Warehousing and distribution.

Storage and transportation of food must be under conditions that will protect against allergen cross-contact and against biological, chemical (including radiological), and physical contamination of food, as well as against deterioration of the food and the container.

§ 117.110 Defect action levels.

a) The manufacturer, processor, packer, and holder of food must at all times utilize quality control operations that reduce natural or unavoidable defects to the lowest level currently feasible.

b) The mixing of a food containing defects at levels that render that food adulterated with another lot of food is not permitted and renders the final food adulterated, regardless of the defect level of the final food. For examples of defect action levels that may render food adulterated, see the Defect Levels Handbook, which is accessible at http://www.fda.gov/pchfrule and at http://www.fda.gov.

Subpart C—Hazard Analysis and Risk-Based Preventive Controls

§ 117.126 Food safety plan.

a) *Requirement for a food safety plan.*
   1) You must prepare, or have prepared, and implement a written food safety plan.
   2) The food safety plan must be prepared, or its preparation overseen, by one or more preventive controls qualified individuals.

b) *Contents of a food safety plan.* The written food safety plan must include:
   1) The written hazard analysis as required by § 117.130(a)(2);
   2) The written preventive controls as required by § 117.135(b);
   3) The written supply-chain program as required by subpart G of this part;
   4) The written recall plan as required by § 117.139(a); and
   5) The written procedures for monitoring the implementation of the preventive controls as required by § 117.145(a)(1);
   6) The written corrective action procedures as required by § 117.150(a)(1); and
   7) The written verification procedures as required by § 117.165(b).

c) *Records.* The food safety plan required by this section is a record that is subject to the requirements of subpart F of this part.

§ 117.130 Hazard analysis.

a) *Requirement for a hazard analysis.*
   1) You must conduct a hazard analysis to identify and evaluate, based on experience, illness data, scientific reports, and other information, known or reasonably foreseeable hazards for each type of food manufactured, processed, packed, or held at your facility to determine whether there are any hazards requiring a preventive control.
   2) The hazard analysis must be written regardless of its outcome.

b) *Hazard identification.* The hazard identification must consider:
   1) Known or reasonably foreseeable hazards that include:
      i) Biological hazards, including microbiological hazards such as parasites, environmental pathogens, and other pathogens;

ii) Chemical hazards, including radiological hazards, substances such as pesticide and drug residues, natural toxins, decomposition, unapproved food or color additives, and food allergens; and
iii) Physical hazards (such as stones, glass, and metal fragments); and
2) Known or reasonably foreseeable hazards that may be present in the food for any of the following reasons:
i) The hazard occurs naturally;
ii) The hazard may be unintentionally introduced; or
iii) The hazard may be intentionally introduced for purposes of economic gain.

c) *Hazard evaluation.*
1) i) The hazard analysis must include an evaluation of the hazards identified in paragraph (b) of this section to assess the severity of the illness or injury if the hazard were to occur and the probability that the hazard will occur in the absence of preventive controls.
ii) The hazard evaluation required by paragraph (c)(1)(i) of this section must include an evaluation of environmental pathogens whenever a ready-to-eat food is exposed to the environment prior to packaging and the packaged food does not receive a treatment or otherwise include a control measure (such as a formulation lethal to the pathogen) that would significantly minimize the pathogen.
2) The hazard evaluation must consider the effect of the following on the safety of the finished food for the intended consumer:
i) The formulation of the food;
ii) The condition, function, and design of the facility and equipment;
iii) Raw materials and other ingredients;
iv) Transportation practices;
v) Manufacturing/processing procedures;
vi) Packaging activities and labeling activities;
vii) Storage and distribution;
viii) Intended or reasonably foreseeable use;
ix) Sanitation, including employee hygiene; and
x) Any other relevant factors, such as the temporal (*e.g.,* weather-related) nature of some hazards (*e.g.,* levels of some natural toxins).

§ 117.135 Preventive controls.

a) 1) You must identify and implement preventive controls to provide assurances that any hazards requiring a preventive control will be significantly minimized or prevented and the food manufactured, processed, packed, or held by your facility will not be adulterated under section 402 of the Federal Food, Drug, and Cosmetic Act or misbranded under section 403(w) of the Federal Food, Drug, and Cosmetic Act.
2) Preventive controls required by paragraph (a)(1) of this section include:

   i) Controls at critical control points (CCPs), if there are any CCPs; and
   ii) Controls, other than those at CCPs, that are also appropriate for food safety.

b) Preventive controls must be written.
c) Preventive controls include, as appropriate to the facility and the food:
   1) *Process controls.* Process controls include procedures, practices, and processes to ensure the control of parameters during operations such as heat processing, acidifying, irradiating, and refrigerating foods. Process controls must include, as appropriate to the nature of the applicable control and its role in the facility's food safety system:
      i) Parameters associated with the control of the hazard; and
      ii) The maximum or minimum value, or combination of values, to which any biological, chemical, or physical parameter must be controlled to significantly minimize or prevent a hazard requiring a process control.
   2) *Food allergen controls.* Food allergen controls include procedures, practices, and processes to control food allergens. Food allergen controls must include those procedures, practices, and processes employed for:
      i) Ensuring protection of food from allergen cross-contact, including during storage, handling, and use; and
      ii) Labeling the finished food, including ensuring that the finished food is not misbranded under section 403(w) of the Federal Food, Drug, and Cosmetic Act.
   3) *Sanitation controls.* Sanitation controls include procedures, practices, and processes to ensure that the facility is maintained in a sanitary condition adequate to significantly minimize or prevent hazards such as environmental pathogens, biological hazards due to employee handling, and food allergen hazards. Sanitation controls must include, as appropriate to the facility and the food, procedures, practices, and processes for the:
      i) Cleanliness of food-contact surfaces, including food-contact surfaces of utensils and equipment;
      ii) Prevention of allergen crosscontact and cross-contamination from insanitary objects and from personnel to food, food packaging material, and other food-contact surfaces and from raw product to processed product.
   4) *Supply-chain controls.* Supplychain controls include the supply-chain program as required by subpart G of this part.
   5) *Recall plan.* Recall plan as required by § 117.139.
   6) *Other controls.* Preventive controls include any other procedures, practices, and processes necessary to satisfy the requirements of paragraph (a) of this section. Examples of other controls include hygiene training and other current good manufacturing practices.

§ 117.136 Circumstances in which the owner, operator, or agent in charge of a manufacturing/processing facility is not required to implement a preventive control.

a) *Circumstances.* If you are a manufacturer/processor, you are not required to implement a preventive control when you identify a hazard requiring a preventive control (identified hazard) and any of the following circumstances apply:
   1) You determine and document that the type of food (*e.g.*, raw agricultural commodities such as cocoa beans, coffee beans, and grains) could not be consumed without application of an appropriate control.
   2) You rely on your customer who is subject to the requirements for hazard analysis and risk-based preventive controls in this subpart C to ensure that the identified hazard will be significantly minimized or prevented and you:
      i) Disclose in documents accompanying the food, in accordance with the practice of the trade, that the food is "not processed to control [identified hazard]"; and
      ii) Annually obtain from your customer written assurance, subject to the requirements of § 117.137, that the customer has established and is following procedures (identified in the written assurance) that will significantly minimize or prevent the identified hazard.
   3) You rely on your customer who is not subject to the requirements for hazard analysis and risk-based preventive controls in this subpart to provide assurance it is manufacturing, processing, or preparing the food in accordance with applicable food safety requirements and you:
      i) Disclose in documents accompanying the food, in accordance with the practice of the trade, that the food is "not processed to control [identified hazard]"; and
      ii) Annually obtain from your customer written assurance that it is manufacturing, processing, or preparing the food in accordance with applicable food safety requirements.
   4) You rely on your customer to provide assurance that the food will be processed to control the identified hazard by an entity in the distribution chain subsequent to the customer and you:
      i) Disclose in documents accompanying the food, in accordance with the practice of the trade, that the food is "not processed to control [identified hazard]"; and
      ii) Annually obtain from your customer written assurance, subject to the requirements of § 117.137, that your customer:
         A) Will disclose in documents accompanying the food, in accordance with the practice of the trade, that the food is "not processed to control [identified hazard]"; and
         B) Will only sell to another entity that agrees, in writing, it will:

1) Follow procedures (identified in a written assurance) that will significantly minimize or prevent the identified hazard (if the entity is subject to the requirements for hazard analysis and risk-based preventive controls in this subpart) or manufacture, process, or prepare the food in accordance with applicable food safety requirements (if the entity is not subject to the requirements for hazard analysis and risk-based preventive controls in this subpart); or

2) Obtain a similar written assurance from the entity's customer, subject to the requirements of § 117.137, as in paragraphs (a)(4)(ii)(A) and (B) of this section, as appropriate; or

5) You have established, documented, and implemented a system that ensures control, at a subsequent distribution step, of the hazards in the food product you distribute and you document the implementation of that system.

b) *Records.* You must document any circumstance, specified in paragraph (a) of this section, that applies to you, including:
   1) A determination, in accordance with paragraph (a) of this section, that the type of food could not be consumed without application of an appropriate control;
   2) The annual written assurance from your customer in accordance with paragraph (a)(2) of this section;
   3) The annual written assurance from your customer in accordance with paragraph (a)(3) of this section;
   4) The annual written assurance from your customer in accordance with paragraph (a)(4) of this section; and
   5) Your system, in accordance with paragraph (a)(5) of this section, that ensures control, at a subsequent distribution step, of the hazards in the food product you distribute.

§ 117.137 Provision of assurances required under § 117.136(a)(2), (3), and (4).

A facility that provides a written assurance under § 117.136(a)(2), (3), or (4) must act consistently with the assurance and document its actions taken to satisfy the written assurance.

§ 117.139 Recall plan.

For food with a hazard requiring a preventive control:

a) You must establish a written recall plan for the food.
b) The written recall plan must include procedures that describe the steps to be taken, and assign responsibility for taking those steps, to perform the following actions as appropriate to the facility:
   1) Directly notify the direct consignees of the food being recalled, including how to return or dispose of the affected food;

2) Notify the public about any hazard presented by the food when appropriate to protect public health;
3) Conduct effectiveness checks to verify that the recall is carried out; and
4) Appropriately dispose of recalled food—*e.g.*, through reprocessing, reworking, diverting to a use that does not present a safety concern, or destroying the food.

§ 117.140 Preventive control management components.

a) Except as provided by paragraphs (b) and (c) of this section, the preventive controls required under § 117.135 are subject to the following preventive control management components as appropriate to ensure the effectiveness of the preventive controls, taking into account the nature of the preventive control and its role in the facility's food safety system:
   1) Monitoring in accordance with § 117.145;
   2) Corrective actions and corrections in accordance with § 117.150; and (3) Verification in accordance with § 117.155.
b) The supply-chain program established in subpart G of this part is subject to the following preventive control management components as appropriate to ensure the effectiveness of the supply-chain program, taking into account the nature of the hazard controlled before receipt of the raw material or other ingredient:
   1) Corrective actions and corrections in accordance with § 117.150, taking into account the nature of any supplier non-conformance;
   2) Review of records in accordance with § 117.165(a)(4); and
   3) Reanalysis in accordance with § 117.170.
c) The recall plan established in § 117.139 is not subject to the requirements of paragraph (a) of this section.

§ 117.145 Monitoring.

As appropriate to the nature of the preventive control and its role in the facility's food safety system:

a) *Written procedures.* You must establish and implement written procedures, including the frequency with which they are to be performed, for monitoring the preventive control; and.
b) *Monitoring.* You must monitor the preventive controls with adequate frequency to provide assurance that they are consistently performed.
c) *Records.*
   1) *Requirement to document monitoring.* You must document the monitoring of preventive controls in accordance with this section in records that are subject to verification in accordance with § 117.155(a)(2) and records review in accordance with § 117.165(a)(4)(i).
   2) *Exception records.*
      i) Records of refrigeration temperature during storage of food that requires time/temperature control to significantly minimize or

prevent the growth of, or toxin production by, pathogens may be affirmative records demonstrating temperature is controlled or exception records demonstrating loss of temperature control.

ii) Exception records may be adequate in circumstances other than monitoring of refrigeration temperature.

§ 117.150 Corrective actions and corrections.

a) *Corrective action procedures.* As appropriate to the nature of the hazard and the nature of the preventive control, except as provided by paragraph (c) of this section:
   1) You must establish and implement written corrective action procedures that must be taken if preventive controls are not properly implemented, including procedures to address, as appropriate:
      i) The presence of a pathogen or appropriate indicator organism in a ready-to-eat product detected as a result of product testing conducted in accordance with § 117.165(a)(2); and
      ii) The presence of an environmental pathogen or appropriate indicator organism detected through the environmental monitoring conducted in accordance with § 117.165(a)(3).
   2) The corrective action procedures must describe the steps to be taken to ensure that:
      i) Appropriate action is taken to identify and correct a problem that has occurred with implementation of a preventive control;
      ii) Appropriate action is taken, when necessary, to reduce the likelihood that the problem will recur;
      iii) All affected food is evaluated for safety; and
      iv) All affected food is prevented from entering into commerce, if you cannot ensure that the affected food is not adulterated under section 402 of the Federal Food, Drug, and Cosmetic Act or misbranded under section 403(w) of the Federal Food, Drug, and Cosmetic Act.

b) *Corrective action in the event of an unanticipated food safety problem.*
   1) Except as provided by paragraph (c) of this section, you are subject to the requirements of paragraphs (b)(2) of this section if any of the following circumstances apply:
      i) A preventive control is not properly implemented and a corrective action procedure has not been established;
      ii) A preventive control, combination of preventive controls, or the food safety plan as a whole is found to be ineffective; or
      iii) A review of records in accordance with § 117.165(a)(4) finds that the records are not complete, the activities conducted did not occur in accordance with the food safety plan, or appropriate decisions were not made about corrective actions.

2) If any of the circumstances listed in paragraph (b)(1) of this section apply, you must:
   i) Take corrective action to identify and correct the problem, reduce the likelihood that the problem will recur, evaluate all affected food for safety, and, as necessary, prevent affected food from entering commerce as would be done following a corrective action procedure under paragraphs (a)(2)(i) through (iv) of this section; and
   ii) When appropriate, reanalyze the food safety plan in accordance with § 117.170 to determine whether modification of the food safety plan is required.

c) *Corrections.* You do not need to comply with the requirements of paragraphs (a) and (b) of this section if:
   1) You take action, in a timely manner, to identify and correct conditions and practices that are not consistent with the food allergen controls in § 117.135(c)(2)(i) or the sanitation controls in § 117.135(c)(3)(i) or (ii); or
   2) You take action, in a timely manner, to identify and correct a minor and isolated problem that does not directly impact product safety.

d) *Records.* All corrective actions (and, when appropriate, corrections) taken in accordance with this section must be documented in records. These records are subject to verification in accordance with § 117.155(a)(3) and records review in accordance with § 117.165(a)(4)(i).

§ 117.155 Verification.

a) *Verification activities.* Verification activities must include, as appropriate to the nature of the preventive control and its role in the facility's food safety system:
   1) Validation in accordance with § 117.160.
   2) Verification that monitoring is being conducted as required by § 117.140 (and in accordance with § 117.145).
   3) Verification that appropriate decisions about corrective actions are being made as required by § 117.140 (and in accordance with § 117.150).
   4) Verification of implementation and effectiveness in accordance with § 117.165; and
   5) Reanalysis in accordance with § 117.170.

b) *Documentation.* All verification activities conducted in accordance with this section must be documented in records.

§ 117.160 Validation.

a) You must validate that the preventive controls identified and implemented in accordance with § 117.135 are adequate to control the hazard as appropriate to the nature of the preventive control and its role in the facility's food safety system.

b) The validation of the preventive controls:

1) Must be performed (or overseen) by a preventive controls qualified individual:
   i) A) Prior to implementation of the food safety plan; or
      B) When necessary to demonstrate the control measures can be implemented as designed:
         1) Within 90 calendar days after production of the applicable food first begins; or
         2) Within a reasonable timeframe, provided that the preventive controls qualified individual prepares (or oversees the preparation of) a written justification for a timeframe that exceeds 90 calendar days after production of the applicable food first begins;
   ii) Whenever a change to a control measure or combination of control measures could impact whether the control measure or combination of control measures, when properly implemented, will effectively control the hazards; and
   iii) Whenever a reanalysis of the food safety plan reveals the need to do so;
2) Must include obtaining and evaluating scientific and technical evidence (or, when such evidence is not available or is inadequate, conducting studies) to determine whether the preventive controls, when properly implemented, will effectively control the hazards; and

c) You do not need to validate:
   1) The food allergen controls in § 117.135(c)(2);
   2) The sanitation controls in § 117.135(c)(3);
   3) The recall plan in § 117.139;
   4) The supply-chain program in subpart G of this part; and
   5) Other preventive controls, if the preventive controls qualified individual prepares (or oversees the preparation of) a written justification that validation is not applicable based on factors such as the nature of the hazard, and the nature of the preventive control and its role in the facility's food safety system.

§ 117.165 Verification of implementation and effectiveness.

a) *Verification activities.* You must verify that the preventive controls are consistently implemented and are effectively and significantly minimizing or preventing the hazards. To do so you must conduct activities that include the following, as appropriate to the facility, the food, and the nature of the preventive control and its role in the facility's food safety system:
   1) Calibration of process monitoring instruments and verification instruments (or checking them for accuracy);
   2) Product testing, for a pathogen (or appropriate indicator organism) or other hazard;

3) Environmental monitoring, for an environmental pathogen or for an appropriate indicator organism, if contamination of a ready-to-eat food with an environmental pathogen is a hazard requiring a preventive control, by collecting and testing environmental samples; and
4) Review of the following records within the specified timeframes, by (or under the oversight of) a preventive controls qualified individual, to ensure that the records are complete, the activities reflected in the records occurred in accordance with the food safety plan, the preventive controls are effective, and appropriate decisions were made about corrective actions:
   i) Records of monitoring and corrective action records within 7 working days after the records are created or within a reasonable timeframe, provided that the preventive controls qualified individual prepares (or oversees the preparation of) a written justification for a timeframe that exceeds 7 working days; and
   ii) Records of calibration, testing (*e.g.*, product testing, environmental monitoring), supplier and supply-chain verification activities, and other verification activities within a reasonable time after the records are created; and
5) Other activities appropriate for verification of implementation and effectiveness.

b) *Written procedures.* As appropriate to the facility, the food, the nature of the preventive control, and the role of the preventive control in the facility's food safety system, you must establish and implement written procedures for the following activities:
   1) The method and frequency of calibrating process monitoring instruments and verification instruments (or checking them for accuracy) as required by paragraph (a)(1) of this section.
   2) Product testing as required by paragraph (a)(2) of this section. Procedures for product testing must:
      i) Be scientifically valid;
      ii) Identify the test microorganism(s) or other analyte(s);
      iii) Specify the procedures for identifying samples, including their relationship to specific lots of product;
      iv) Include the procedures for sampling, including the number of samples and the sampling frequency;
      v) Identify the test(s) conducted, including the analytical method(s) used;
      vi) Identify the laboratory conducting the testing; and
      vii) Include the corrective action procedures required by § 117.150(a)(1).
   3) Environmental monitoring as required by paragraph (a)(3) of this section. Procedures for environmental monitoring must:
      i) Be scientifically valid;

ii) Identify the test microorganism(s);
iii) Identify the locations from which samples will be collected and the number of sites to be tested during routine environmental monitoring. The number and location of sampling sites must be adequate to determine whether preventive controls are effective;
iv) Identify the timing and frequency for collecting and testing samples. The timing and frequency for collecting and testing samples must be adequate to determine whether preventive controls are effective;
v) Identify the test(s) conducted, including the analytical method(s) used;
vi) Identify the laboratory conducting the testing; and
vii) Include the corrective action procedures required by § 117.150(a)(1).

§ 117.170 Reanalysis.

a) You must conduct a reanalysis of the food safety plan as a whole at least once every 3 years;
b) You must conduct a reanalysis of the food safety plan as a whole, or the applicable portion of the food safety plan:
   1) Whenever a significant change in the activities conducted at your facility creates a reasonable potential for a new hazard or creates a significant increase in a previously identified hazard;
   2) Whenever you become aware of new information about potential hazards associated with the food;
   3) Whenever appropriate after an unanticipated food safety problem in accordance with § 117.150(b); and
   4) Whenever you find that a preventive control, combination of preventive controls, or the food safety plan as a whole is ineffective.
c) You must complete the reanalysis required by paragraphs (a) and (b) of this section and validate, as appropriate to the nature of the preventive control and its role in the facility's food safety system, any additional preventive controls needed to address the hazard identified:
   1) Before any change in activities (including any change in preventive control) at the facility is operative; or
   2) When necessary to demonstrate the control measures can be implemented as designed:
      i) Within 90 calendar days after production of the applicable food first begins; or
      ii) Within a reasonable timeframe, provided that the preventive controls qualified individual prepares (or oversees the preparation of) a written justification for a timeframe that exceeds 90-calendar days after production of the applicable food first begins.
d) You must revise the written food safety plan if a significant change in the activities conducted at your facility creates a reasonable potential for a new

hazard or a significant increase in a previously identified hazard or document the basis for the conclusion that no revisions are needed.

e) A preventive controls qualified individual must perform (or oversee) the reanalysis.

f) You must conduct a reanalysis of the food safety plan when FDA determines it is necessary to respond to new hazards and developments in scientific understanding.

§ 117.180 Requirements applicable to a preventive controls qualified individual and a qualified auditor.

a) One or more preventive controls qualified individuals must do or oversee the following:
   1) Preparation of the food safety plan (§ 117.126(a)(2));
   2) Validation of the preventive controls (§ 117.160(b)(1));
   3) Written justification for validation to be performed in a timeframe that exceeds the first 90 calendar days of production of the applicable food;
   4) Determination that validation is not required (§ 117.160(c)(5));
   5) Review of records (§ 117.165(a)(4));
   6) Written justification for review of records of monitoring and corrective actions within a timeframe that exceeds 7 working days;
   7) Reanalysis of the food safety plan (§ 117.170(d)); and
   8) Determination that reanalysis can be completed, and additional preventive controls validated, as appropriate to the nature of the preventive control and its role in the facility's food safety system, in a timeframe that exceeds the first 90 calendar days of production of the applicable food.

b) A qualified auditor must conduct an onsite audit (§ 117.435(a)).

c) 1) To be a preventive controls qualified individual, the individual must have successfully completed training in the development and application of risk-based preventive controls at least equivalent to that received under a standardized curriculum recognized as adequate by FDA or be otherwise qualified through job experience to develop and apply a food safety system. Job experience may qualify an individual to perform these functions if such experience has provided an individual with knowledge at least equivalent to that provided through the standardized curriculum. This individual may be, but is not required to be, an employee of the facility.
   2) To be a qualified auditor, a qualified individual must have technical expertise obtained through education, training, or experience (or a combination thereof) necessary to perform the auditing function.

d) All applicable training in the development and application of riskbased preventive controls must be documented in records, including the date of the training, the type of training, and the person(s) trained.

§ 117.190 Implementation records required for this subpart.

a) You must establish and maintain the following records documenting implementation of the food safety plan:
   1) Documentation, as required by § 117.136(b), of the basis for not establishing a preventive control in accordance with § 117.136(a);
   2) Records that document the monitoring of preventive controls;
   3) Records that document corrective actions;
   4) Records that document verification, including, as applicable, those related to:
      i) Validation;
      ii) Verification of monitoring;
      iii) Verification of corrective actions;
      iv) Calibration of process monitoring and verification instruments;
      v) Product testing;
      vi) Environmental monitoring;
      vii) Records review; and
      viii) Reanalysis;
   5) Records that document the supplychain program; and
   6) Records that document applicable training for the preventive controls qualified individual and the qualified auditor.

b) The records that you must establish and maintain are subject to the requirements of subpart F of this part.

Subpart D—Modified Requirements

§ 117.201 Modified requirements that apply to a qualified facility.

a) *Attestations to be submitted.* A qualified facility must submit the following attestations to FDA:
   1) An attestation that the facility is a qualified facility as defined in § 117.3. For the purpose of determining whether a facility satisfies the definition of qualified facility, the baseline year for calculating the adjustment for inflation is 2011; and
   2) i) An attestation that you have identified the potential hazards associated with the food being produced, are implementing preventive controls to address the hazards, and are monitoring the performance of the preventive controls to ensure that such controls are effective; or
      ii) An attestation that the facility is in compliance with State, local, county, tribal, or other applicable non-Federal food safety law, including relevant laws and regulations of foreign countries, including an attestation based on licenses, inspection reports, certificates, permits, credentials, certification by an appropriate agency (such as a State department of agriculture), or other evidence of oversight.

b) *Procedure for submission.* The attestations required by paragraph (a) of this section must be submitted to FDA by one of the following means:
   1) *Electronic submission.* To submit electronically, go to http://www.fda.gov/ furls and follow the instructions. This Web site is available from wherever the Internet is accessible, including libraries, copy centers, schools, and Internet cafes. FDA encourages electronic submission.
   2) *Submission by mail.*
      i) You must use Form FDA 3942a. You may obtain a copy of this form by any of the following mechanisms:
         A) Download it from http://www.fda.gov/pchfrule;
         B) Write to the U.S. Food and Drug Administration (HFS–681), 5100 Paint Branch Parkway, College Park, MD 20550; or
         C) Request a copy of this form by phone at 1–800–216–7331 or 301–575–0156.
      ii) Send a paper Form FDA 3942a to the U.S. Food and Drug Administration (HFS–681), 5100 Paint Branch Parkway, College Park, MD 20550. We recommend that you submit a paper copy only if your facility does not have reasonable access to the Internet.

c) *Frequency of determination of status and submission.*
   1) A facility must determine and document its status as a qualified facility on an annual basis no later than July 1 of each calendar year.
   2) The attestations required by paragraph (a) of this section must be:
      i) Submitted to FDA initially:
         A) By December 17, 2018, for a facility that begins manufacturing, processing, packing, or holding food before September 17, 2018;
         B) Before beginning operations, for a facility that begins manufacturing, processing, packing, or holding food after September 17, 2018; or
         C) By July 31 of the applicable calendar year, when the status of a facility changes from "not a qualified facility" to "qualified facility" based on the annual determination required by paragraph (c)(1) of this section; and
      ii) Beginning in 2020, submitted to FDA every 2 years during the period beginning on October 1 and ending on December 31.
   3) When the status of a facility changes from "qualified facility" to "not a qualified facility" based on the annual determination required by paragraph (c)(1) of this section, the facility must notify FDA of that change in status using Form 3942a by July 31 of the applicable calendar year.

d) *Timeframe for compliance with subparts C and G of this part when the facility status changes to "not a qualified facility."* When the status of a facility changes from "qualified facility" to "not a qualified facility," the facility must comply with subparts C and G of this part no later than December 31 of the applicable calendar year unless otherwise agreed to by FDA and the facility.

e) *Notification to consumers.* A qualified facility that does not submit attestations under paragraph (a)(2)(i) of this section must provide notification to consumers as to the name and complete business address of the facility where the food was manufactured or processed (including the street address or P.O. box, city, state, and zip code for domestic facilities, and comparable full address information for foreign facilities), as follows:
   1) If a food packaging label is required, the notification required by paragraph (e) of this section must appear prominently and conspicuously on the label of the food.
   2) If a food packaging label is not required, the notification required by paragraph (e) of this section must appear prominently and conspicuously, at the point of purchase, on a label, poster, sign, placard, or documents delivered contemporaneously with the food in the normal course of business, or in an electronic notice, in the case of Internet sales.

f) *Records.*
   1) A qualified facility must maintain those records relied upon to support the attestations that are required by paragraph (a) of this section.
   2) The records that a qualified facility must maintain are subject to the requirements of subpart F of this part.

§ 117.206 Modified requirements that apply to a facility solely engaged in the storage of unexposed packaged food.

a) If a facility that is solely engaged in the storage of unexposed packaged food stores any such refrigerated packaged food that requires time/ temperature control to significantly minimize or prevent the growth of, or toxin production by pathogens, the facility must conduct the following activities as appropriate to ensure the effectiveness of the temperature controls:
   1) Establish and implement temperature controls adequate to significantly minimize or prevent the growth of, or toxin production by, pathogens;
   2) Monitor the temperature controls with adequate frequency to provide assurance that the temperature controls are consistently performed;
   3) If there is a loss of temperature control that may impact the safety of such refrigerated packaged food, take appropriate corrective actions to:
      i) Correct the problem and reduce the likelihood that the problem will recur;
      ii) Evaluate all affected food for safety; and
      iii) Prevent the food from entering commerce, if you cannot ensure the affected food is not adulterated under section 402 of the Federal Food, Drug, and Cosmetic Act;
   4) Verify that temperature controls are consistently implemented by:
      i) Calibrating temperature monitoring and recording devices (or checking them for accuracy);
      ii) Reviewing records of calibration within a reasonable time after the records are created; and

iii) Reviewing records of monitoring and corrective actions taken to correct a problem with the control of temperature within 7 working days after the records are created or within a reasonable timeframe, provided that the preventive controls qualified individual prepares (or oversees the preparation of) a written justification for a timeframe that exceeds 7 working days;

5) Establish and maintain the following records:
   i) Records (whether affirmative records demonstrating temperature is controlled or exception records demonstrating loss of temperature control) documenting the monitoring of temperature controls for any such refrigerated packaged food;
   ii) Records of corrective actions taken when there is a loss of temperature control that may impact the safety of any such refrigerated packaged food; and
   iii) Records documenting verification activities.

b) The records that a facility must establish and maintain under paragraph (a)(5) of this section are subject to the requirements of subpart F of this part.

Subpart E—Withdrawal of a Qualified Facility Exemption

§ 117.251 Circumstances that may lead FDA to withdraw a qualified facility exemption.

a) FDA may withdraw a qualified facility exemption under § 117.5(a):
   1) In the event of an active investigation of a foodborne illness outbreak that is directly linked to the qualified facility; or
   2) If FDA determines that it is necessary to protect the public health and prevent or mitigate a foodborne illness outbreak based on conditions or conduct associated with the qualified facility that are material to the safety of the food manufactured, processed, packed, or held at such facility.

b) Before FDA issues an order to withdraw a qualified facility exemption, FDA:
   1) May consider one or more other actions to protect the public health or mitigate a foodborne illness outbreak, including a warning letter, recall, administrative detention, suspension of registration, refusal of food offered for import, seizure, and injunction;
   2) Must notify the owner, operator, or agent in charge of the facility, in writing, of circumstances that may lead FDA to withdraw the exemption, and provide an opportunity for the owner, operator, or agent in charge of the facility to respond in writing, within 15 calendar days of the date of receipt of the notification, to FDA's notification; and
   3) Must consider the actions taken by the facility to address the circumstances that may lead FDA to withdraw the exemption.

§ 117.254 Issuance of an order to withdraw a qualified facility exemption.

a) An FDA District Director in whose district the qualified facility is located (or, in the case of a foreign facility, the Director of the Office of Compliance in the Center for Food Safety and Applied Nutrition), or an FDA official senior to either such Director, must approve an order to withdraw the exemption before the order is issued.
b) Any officer or qualified employee of FDA may issue an order to withdraw the exemption after it has been approved in accordance with paragraph (a) of this section.
c) FDA must issue an order to withdraw the exemption to the owner, operator, or agent in charge of the facility.
d) FDA must issue an order to withdraw the exemption in writing, signed and dated by the officer or qualified employee of FDA who is issuing the order.

§ 117.257 Contents of an order to withdraw a qualified facility exemption.

An order to withdraw a qualified facility exemption under § 117.5(a) must include the following information:

a) The date of the order;
b) The name, address, and location of the qualified facility;
c) A brief, general statement of the reasons for the order, including information relevant to one or both of the following circumstances that leads FDA to issue the order:
    1) An active investigation of a foodborne illness outbreak that is directly linked to the facility; or
    2) Conditions or conduct associated with a qualified facility that are material to the safety of the food manufactured, processed, packed, or held at such facility.
d) A statement that the facility must either:
    1) Comply with subparts C and G of this part on the date that is 120 calendar days after the date of receipt of the order, or within a reasonable timeframe, agreed to by FDA, based on a written justification, submitted to FDA, for a timeframe that exceeds 120 calendar days from the date of receipt of the order; or
    2) Appeal the order within 15 calendar days of the date of receipt of the order in accordance with the requirements of § 117.264.
e) A statement that a facility may request that FDA reinstate an exemption that was withdrawn by following the procedures in § 117.287.
f) The text of section 418(l) of the Federal Food, Drug, and Cosmetic Act and of this subpart;
g) A statement that any informal hearing on an appeal of the order must be conducted as a regulatory hearing under part 16 of this chapter, with certain exceptions described in § 117.270;

h) The mailing address, telephone number, email address, and facsimile number of the FDA district office and the name of the FDA District Director in whose district the facility is located (or, in the case of a foreign facility, the same information for the Director of the Office of Compliance in the Center for Food Safety and Applied Nutrition); and
i) The name and the title of the FDA representative who approved the order.

§ 117.260 Compliance with, or appeal of, an order to withdraw a qualified facility exemption.

a) If you receive an order under § 117.254 to withdraw a qualified facility exemption, you must either:
   1) Comply with applicable requirements of this part within 120 calendar days of the date of receipt of the order, or within a reasonable timeframe, agreed to by FDA, based on a written justification, submitted to FDA, for a timeframe that exceeds 120 calendar days from the date of receipt of the order; or
   2) Appeal the order within 15 calendar days of the date of receipt of the order in accordance with the requirements of § 117.264.
b) Submission of an appeal, including submission of a request for an informal hearing, will not operate to delay or stay any administrative action, including enforcement action by FDA, unless the Commissioner of Food and Drugs, as a matter of discretion, determines that delay or a stay is in the public interest.
c) If you appeal the order, and FDA confirms the order:
   1) You must comply with applicable requirements of this part within 120 calendar days of the date of receipt of the order, or within a reasonable timeframe, agreed to by FDA, based on a written justification, submitted to FDA, for a timeframe that exceeds 120 calendar days from the date of receipt of the order; and
   2) You are no longer subject to the modified requirements in § 117.201.

§ 117.264 Procedure for submitting an appeal.

a) To appeal an order to withdraw a qualified facility exemption, you must:
   1) Submit the appeal in writing to the FDA District Director in whose district the facility is located (or, in the case of a foreign facility, the Director of the Office of Compliance in the Center for Food Safety and Applied Nutrition), at the mailing address, email address, or facsimile number identified in the order within 15 calendar days of the date of receipt of confirmation of the order;
   2) Respond with particularity to the facts and issues contained in the order, including any supporting documentation upon which you rely.
b) In a written appeal of the order withdrawing an exemption provided under § 117.5(a), you may include a written request for an informal hearing as provided in § 117.267.

§ 117.267 Procedure for requesting an informal hearing.

a) If you appeal the order, you:
   1) May request an informal hearing; and
   2) Must submit any request for an informal hearing together with your written appeal submitted in accordance with § 117.264 within 15 calendar days of the date of receipt of the order.

b) A request for an informal hearing may be denied, in whole or in part, if the presiding officer determines that no genuine and substantial issue of material fact has been raised by the material submitted. If the presiding officer determines that a hearing is not justified, written notice of the determination will be given to you explaining the reason for the denial.

§ 117.270 Requirements applicable to an informal hearing.

If you request an informal hearing, and FDA grants the request:

a) The hearing will be held within 15 calendar days after the date the appeal is filed or, if applicable, within a timeframe agreed upon in writing by you and FDA.

b) The presiding officer may require that a hearing conducted under this subpartbecompletedwithin1-calendarday,asappropriate.DSK3SPTVN1PROD with RULES2

c) FDA must conduct the hearing in accordance with part 16 of this chapter, except that:
   1) The order withdrawing an exemption under §§ 117.254 and 117.257, rather than the notice under § 16.22(a) of this chapter, provides notice of opportunity for a hearing under this section and is part of the administrative record of the regulatory hearing under § 16.80(a) of this chapter.
   2) A request for a hearing under this subpart must be addressed to the FDA District Director (or, in the case of a foreign facility, the Director of the Office of Compliance in the Center for Food Safety and Applied Nutrition) as provided in the order withdrawing an exemption.
   3) Section 117.274, rather than § 16.42(a) of this chapter, describes the FDA employees who preside at hearings under this subpart.
   4) Section 16.60(e) and (f) of this chapter does not apply to a hearing under this subpart. The presiding officer must prepare a written report of the hearing. All written material presented at the hearing will be attached to the report. The presiding officer must include as part of the report of the hearing a finding on the credibility of witnesses (other than expert witnesses) whenever credibility is a material issue, and must include a proposed decision, with a statement of reasons. The hearing participant may review and comment on the presiding officer's report within 2- calendar days of issuance of the report. The presiding officer will then issue the final decision.
   5) Section 16.80(a)(4) of this chapter does not apply to a regulatory hearing under this subpart. The presiding officer's report of the hearing and any

comments on the report by the hearing participant under § 117.270(c)(4) are part of the administrative record.

6) No party shall have the right, under § 16.119 of this chapter to petition the Commissioner of Food and Drugs for reconsideration or a stay of the presiding officer's final decision.
7) If FDA grants a request for an informal hearing on an appeal of an order withdrawing an exemption, the hearing must be conducted as a regulatory hearing under a regulation in accordance with part 16 of this chapter, except that § 16.95(b) of this chapter does not apply to a hearing under this subpart. With respect to a regulatory hearing under this subpart, the administrative record of the hearing specified in §§ 16.80(a)(1) through (3) and (a)(5) of this chapter and 117.270(c)(5) constitutes the exclusive record for the presiding officer's final decision. For purposes of judicial review under § 10.45 of this chapter, the record of the administrative proceeding consists of the record of the hearing and the presiding officer's final decision.

§ 117.274 Presiding officer for an appeal and for an informal hearing.

The presiding officer for an appeal, and for an informal hearing, must be an FDA Regional Food and Drug Director or another FDA official senior to an FDA District Director.

§ 117.277 Timeframe for issuing a decision on an appeal.

a) If you appeal the order without requesting a hearing, the presiding officer must issue a written report that includes a final decision confirming or revoking the withdrawal by the 10th calendar day after the appeal is filed.
b) If you appeal the order and request an informal hearing:
    1) If FDA grants the request for a hearing and the hearing is held, the presiding officer must provide a 2- calendar day opportunity for the hearing participants to review and submit comments on the report of the hearing under § 117.270(c)(4), and must issue a final decision within 10-calendar days after the hearing is held; or
    2) If FDA denies the request for a hearing, the presiding officer must issue a final decision on the appeal confirming or revoking the withdrawal within 10 calendar days after the date the appeal is filed.

§ 117.280 Revocation of an order to withdraw a qualified facility exemption.

An order to withdraw a qualified facility exemption is revoked if:

a) You appeal the order and request an informal hearing, FDA grants the request for an informal hearing, and the presiding officer does not confirm the order within the 10-calendar days after the hearing, or issues a decision revoking the order within that time; or

b) You appeal the order and request an informal hearing, FDA denies the request for an informal hearing, and FDA does not confirm the order within the 10-calendar days after the appeal is filed, or issues a decision revoking the order within that time; or
c) You appeal the order without requesting an informal hearing, and FDA does not confirm the order within the 10-calendar days after the appeal is filed, or issues a decision revoking the order within that time.

§ 117.284 Final agency action.
Confirmation of a withdrawal order by the presiding officer is considered a final agency action for purposes of 5 U.S.C. 702.

§ 117.287 Reinstatement of a qualified facility exemption that was withdrawn.

a) If the FDA District Director in whose district your facility is located (or, in the case of a foreign facility, the Director of the Office of Compliance in the Center for Food Safety and Applied Nutrition) determines that a facility has adequately resolved any problems with the conditions and conduct that are material to the safety of the food manufactured, processed, packed, or held at the facility and that continued withdrawal of the exemption is not necessary to protect public health and prevent or mitigate a foodborne illness outbreak, the FDA District Director in whose district your facility is located (or, in the case of a foreign facility, the Director of the Office of Compliance in the Center for Food Safety and Applied Nutrition) will, on his own initiative or on the request of a facility, reinstate the exemption.
b) You may ask FDA to reinstate an exemption that has been withdrawn under the procedures of this subpart as follows:
   1) Submit a request, in writing, to the FDA District Director in whose district your facility is located (or, in the case of a foreign facility, the Director of the Office of Compliance in the Center for Food Safety and Applied Nutrition); and
   2) Present data and information to demonstrate that you have adequately resolved any problems with the conditions and conduct that are material to the safety of the food manufactured, processed, packed, or held at your facility, such that continued withdrawal of the exemption is not necessary to protect public health and prevent or mitigate a foodborne illness outbreak.
c) If your exemption was withdrawn under § 117.251(a)(1) and FDA later determines, after finishing the active investigation of a foodborne illness outbreak, that the outbreak is not directly linked to your facility, FDA will reinstate your exemption under § 117.5(a), and FDA will notify you in writing that your exempt status has been reinstated.
d) If your exemption was withdrawn under both § 117.251(a)(1) and (2) and FDA later determines, after finishing the active investigation of a foodborne

illness outbreak, that the outbreak is not directly linked to your facility, FDA will inform you of this finding, and you may ask FDA to reinstate your exemption under § 117.5(a) in accordance with the requirements of paragraph (b) of this section.

Subpart F—Requirements Applying to Records That Must Be Established and Maintained

§ 117.301 Records subject to the requirements of this subpart.

a) Except as provided by paragraphs (b) and (c) of this section, all records required by this part are subject to all requirements of this subpart.
b) The requirements of § 117.310 apply only to the written food safety plan.
c) The requirements of § 117.305(b),
d) (d), (e), and (f) do not apply to the records required by § 117.201.

§ 117.305 General requirements applying to records.

Records must:

a) Be kept as original records, true copies (such as photocopies, pictures, scanned copies, microfilm, microfiche, or other accurate reproductions of the original records), or electronic records;
b) Contain the actual values and observations obtained during monitoring and, as appropriate, during verification activities;
c) Be accurate, indelible, and legible;
d) Be created concurrently with performance of the activity documented;
e) Be as detailed as necessary to provide history of work performed; and
f) Include:
   1) Information adequate to identify the plant or facility (*e.g.*, the name, and when necessary, the location of the plant or facility);
   2) The date and, when appropriate, the time of the activity documented;
   3) The signature or initials of the person performing the activity; and
   4) Where appropriate, the identity of the product and the lot code, if any.
g) Records that are established or maintained to satisfy the requirements of this part and that meet the definition of electronic records in § 11.3(b)(6) of this chapter are exempt from the requirements of part 11 of this chapter. Records that satisfy the requirements of this part, but that also are required under other applicable statutory provisions or regulations, remain subject to part 11 of this chapter.

§ 117.310 Additional requirements applying to the food safety plan.

The owner, operator, or agent in charge of the facility must sign and date the food safety plan:

a) Upon initial completion; and
b) Upon any modification.

§ 117.315 Requirements for record retention.

a) 1) All records required by this part must be retained at the plant or facility for at least 2 years after the date they were prepared.

2) Records that a facility relies on during the 3-year period preceding the applicable calendar year to support its status as a qualified facility must be retained at the facility as long as necessary to support the status of a facility as a qualified facility during the applicable calendar year.

b) Records that relate to the general adequacy of the equipment or processes being used by a facility, including the results of scientific studies and evaluations, must be retained by the facility for at least 2 years after their use is discontinued (*e.g.,* because the facility has updated the written food safety plan (§ 117.126) or records that document validation of the written food safety plan (§ 117.155(b)));

c) Except for the food safety plan, offsite storage of records is permitted if such records can be retrieved and provided onsite within 24 hours of request for official review. The food safety plan must remain onsite. Electronic records are considered to be onsite if they are accessible from an onsite location.

d) If the plant or facility is closed for a prolonged period, the food safety plan may be transferred to some other reasonably accessible location but must be returned to the plant or facility within 24 hours for official review upon request.

§ 117.320 Requirements for official review.

All records required by this part must be made promptly available to a duly authorized representative of the Secretary of Health and Human Services for official review and copying upon oral or written request.

§ 117.325 Public disclosure.

Records obtained by FDA in accordance with this part are subject to the disclosure requirements under part 20 of this chapter.

§ 117.330 Use of existing records.

a) Existing records (*e.g.,* records that are kept to comply with other Federal, State, or local regulations, or for any other reason) do not need to be duplicated if they contain all of the required information and satisfy the requirements of this subpart. Existing records may be supplemented as necessary to include all of the required information and satisfy the requirements of this subpart.

b) The information required by this part does not need to be kept in one set of records. If existing records contain some of the required information, any new information required by this part may be kept either separately or combined with the existing records.

§ 117.335 Special requirements applicable to a written assurance.

a) Any written assurance required by this part must contain the following elements:
   1) Effective date;
   2) Printed names and signatures of authorized officials;
   3) The applicable assurance under:
      i) Section 117.136(a)(2);
      ii) Section 117.136(a)(3);
      iii) Section 117.136(a)(4);
      iv) Section 117.430(c)(2);
      v) Section 117.430(d)(2); or
      vi) Section 117.430(e)(2);

b) A written assurance required under § 117.136(a)(2), (3), or (4) must include:
   1) Acknowledgement that the facility that provides the written assurance assumes legal responsibility to act consistently with the assurance and document its actions taken to satisfy the written assurance; and
   2) Provision that if the assurance is terminated in writing by either entity, responsibility for compliance with the applicable provisions of this part reverts to the manufacturer/processor as of the date of termination.

Subpart G—Supply-Chain Program

§ 117.405 Requirement to establish and implement a supply-chain program.

a) 1) Except as provided by paragraphs (a)(2) and (3) of this section, the receiving facility must establish and implement a risk-based supply-chain program for those raw materials and other ingredients for which the receiving facility has identified a hazard requiring a supply-chain-applied control.
   2) A receiving facility that is an importer, is in compliance with the foreign supplier verification program requirements under part 1, subpart L of this chapter, and has documentation of verification activities conducted under § 1.506(e) of this chapter (which provides assurance that the hazards requiring a supply-chain-applied control for the raw material or other ingredient have been significantly minimized or prevented) need not conduct supplier verification activities for that raw material or other ingredient.
   3) The requirements in this subpart do not apply to food that is supplied for research or evaluation use, provided that such food:
      i) Is not intended for retail sale and is not sold or distributed to the public;
      ii) Is labeled with the statement "Food for research or evaluation use";
      iii) Is supplied in a small quantity that is consistent with a research, analysis, or quality assurance purpose, the food is used only for this purpose, and any unused quantity is properly disposed of; and

iv) Is accompanied with documents, in accordance with the practice of the trade, stating that the food will be used for research or evaluation purposes and cannot be sold or distributed to the public.

b) The supply-chain program must be written.

c) When a supply-chain-applied control is applied by an entity other than the receiving facility's supplier (*e.g.*, when a non-supplier applies controls to certain produce (*i.e.*, produce covered by part 112 of this chapter)), because growing, harvesting, and packing activities are under different management), the receiving facility must:
   1) Verify the supply-chain-applied control; or
   2) Obtain documentation of an appropriate verification activity from another entity, review and assess the entity's applicable documentation, and document that review and assessment.

§ 117.410 General requirements applicable to a supply-chain program.

a) The supply-chain program must include:
   1) Using approved suppliers as required by § 117.420;
   2) Determining appropriate supplier verification activities (including determining the frequency of conducting the activity) as required by § 117.425;
   3) Conducting supplier verification activities as required by §§ 117.430 and 117.435;
   4) Documenting supplier verification activities as required by § 117.475; and
   5) When applicable, verifying a supply-chain-applied control applied by an entity other than the receiving facility's supplier and documenting that verification as required by § 117.475, or obtaining documentation of an appropriate verification activity from another entity, reviewing and assessing that documentation, and documenting the review and assessment as required by § 117.475.

b) The following are appropriate supplier verification activities for raw materials and other ingredients:
   1) Onsite audits;
   2) Sampling and testing of the raw material or other ingredient;
   3) Review of the supplier's relevant food safety records; and
   4) Other appropriate supplier verification activities based on supplier performance and the risk associated with the raw material or other ingredient.

c) The supply-chain program must provide assurance that a hazard requiring a supply-chain-applied control has been significantly minimized or prevented.

d) 1) Except as provided by paragraph (d)(2) of this section, in approving suppliers and determining the appropriate supplier verification activities and the frequency with which they are conducted, the following must be considered:

i) The hazard analysis of the food, including the nature of the hazard controlled before receipt of the raw material or other ingredient, applicable to the raw material and other ingredients;
ii) The entity or entities that will be applying controls for the hazards requiring a supply-chain-applied control;
iii) Supplier performance, including:
    A) The supplier's procedures, processes, and practices related to the safety of the raw material and other ingredients;
    B) Applicable FDA food safety regulations and information relevant to the supplier's compliance with those regulations, including an FDA warning letter or import alert relating to the safety of food and other FDA compliance actions related to food safety (or, when applicable, relevant laws and regulations of a country whose food safety system FDA has officially recognized as comparable or has determined to be equivalent to that of the United States, and information relevant to the supplier's compliance with those laws and regulations); and
    C) The supplier's food safety history relevant to the raw materials or other ingredients that the receiving facility receives from the supplier, including available information about results from testing raw materials or other ingredients for hazards, audit results relating to the safety of the food, and responsiveness of the supplier in correcting problems; and
iv) Any other factors as appropriate and necessary, such as storage and transportation practices.

2) Considering supplier performance can be limited to the supplier's compliance history as required by paragraph (d)(1)(iii)(B) of this section, if the supplier is:
    i) A qualified facility as defined by § 117.3;
    ii) A farm that grows produce and is not a covered farm under part 112 of this chapter in accordance with § 112.4(a), or in accordance with §§ 112.4(b) and 112.5; or
    iii) A shell egg producer that is not subject to the requirements of part 118 of this chapter because it has less than 3,000 laying hens.

e) If the owner, operator, or agent in charge of a receiving facility determines through auditing, verification testing, document review, relevant consumer, customer or other complaints, or otherwise that the supplier is not controlling hazards that the receiving facility has identified as requiring a supply-chain-applied control, the receiving facility must take and document prompt action in accordance with § 117.150 to ensure that raw materials or other ingredients from the supplier do not cause food that is manufactured or processed by the receiving facility to be adulterated under section 402 of the Federal Food, Drug, and Cosmetic Act or misbranded under section 403(w) of the Federal Food, Drug, and Cosmetic Act.

§ 117.415 Responsibilities of the receiving facility.

a) 1) The receiving facility must approve suppliers.
   2) Except as provided by paragraphs (a)(3) and (4) of this section, the receiving facility must determine and conduct appropriate supplier verification activities, and satisfy all documentation requirements of this subpart.
   3) An entity other than the receiving facility may do any of the following, provided that the receiving facility reviews and assesses the entity's applicable documentation, and documents that review and assessment:
      i) Establish written procedures for receiving raw materials and other ingredients by the entity;
      ii) Document that written procedures for receiving raw materials and other ingredients are being followed by the entity; and
      iii) Determine, conduct, or both determine and conduct the appropriate supplier verification activities, with appropriate documentation.
   4) The supplier may conduct and document sampling and testing of raw materials and other ingredients, for the hazard controlled by the supplier, as a supplier verification activity for a particular lot of product and provide such documentation to the receiving facility, provided that the receiving facility reviews and assesses that documentation, and documents that review and assessment.

b) For the purposes of this subpart, a receiving facility may not accept any of the following as a supplier verification activity:
   1) A determination by its supplier of the appropriate supplier verification activities for that supplier;
   2) An audit conducted by its supplier;
   3) A review by its supplier of that supplier's own relevant food safety records; or
   4) The conduct by its supplier of other appropriate supplier verification activities for that supplier within the meaning of § 117.410(b)(4).

c) The requirements of this section do not prohibit a receiving facility from relying on an audit provided by its supplier when the audit of the supplier was conducted by a third-party qualified auditor in accordance with §§ 117.430(f) and 117.435.

§ 117.420 Using approved suppliers.

a) *Approval of suppliers.* The receiving facility must approve suppliers in accordance with the requirements of § 117.410(d), and document that approval, before receiving raw materials and other ingredients received from those suppliers;

b) *Written procedures for receiving raw materials and other ingredients.*
   1) Written procedures for receiving raw materials and other ingredients must be established and followed;
   2) The written procedures for receiving raw materials and other ingredients must ensure that raw materials and other ingredients are received only from approved suppliers (or, when necessary and appropriate, on a temporary basis from unapproved suppliers whose raw materials or other ingredients are subjected to adequate verification activities before acceptance for use); and
   3) Use of the written procedures for receiving raw materials and other ingredients must be documented.

§ 117.425 Determining appropriate supplier verification activities (including determining the frequency of conducting the activity).

Appropriate supplier verification activities (including the frequency of conducting the activity) must be determined in accordance with the requirements of § 117.410(d).

§ 117.430 Conducting supplier verification activities for raw materials and other ingredients.

a) Except as provided by paragraph (c), (d), or (e) of this section, one or more of the supplier verification activities specified in § 117.410(b), as determined under § 117.410(d), must be conducted for each supplier before using the raw material or other ingredient from that supplier and periodically thereafter.
b) 1) Except as provided by paragraph (b)(2) of this section, when a hazard in a raw material or other ingredient will be controlled by the supplier and is one for which there is a reasonable probability that exposure to the hazard will result in serious adverse health consequences or death to humans:
      i) The appropriate supplier verification activity is an onsite audit of the supplier; and
      ii) The audit must be conducted before using the raw material or other ingredient from the supplier and at least annually thereafter.
   2) The requirements of paragraph (b)(1) of this section do not apply if there is a written determination that other verification activities and/or less frequent onsite auditing of the supplier provide adequate assurance that the hazards are controlled.
c) If a supplier is a qualified facility as defined by § 117.3, the receiving facility does not need to comply with paragraphs (a) and (b) of this section if the receiving facility:
   1) Obtains written assurance that the supplier is a qualified facility as defined by § 117.3:

i) Before first approving the supplier for an applicable calendar year; and
ii) On an annual basis thereafter, by December 31 of each calendar year, for the following calendar year; and

2) Obtains written assurance, at least every 2 years, that the supplier is producing the raw material or other ingredient in compliance with applicable FDA food safety regulations (or, when applicable, relevant laws and regulations of a country whose food safety system FDA has officially recognized as comparable or has determined to be equivalent to that of the United States). The written assurance must include either:
   i) A brief description of the preventive controls that the supplier is implementing to control the applicable hazard in the food; or
   ii) A statement that the facility is in compliance with State, local, county, tribal, or other applicable non-Federal food safety law, including relevant laws and regulations of foreign countries.

d) If a supplier is a farm that grows produce and is not a covered farm under part 112 of this chapter in accordance with § 112.4(a), or in accordance with §§ 112.4(b) and 112.5, the receiving facility does not need to comply with paragraphs (a) and (b) of this section for produce that the receiving facility receives from the farm as a raw material or other ingredient if the receiving facility:
   1) Obtains written assurance that the raw material or other ingredient provided by the supplier is not subject to part 112 of this chapter in accordance with § 112.4(a), or in accordance with §§ 112.4(b) and 112.5:
      i) Before first approving the supplier for an applicable calendar year; and
      ii) On an annual basis thereafter, by December 31 of each calendar year, for the following calendar year; and
   2) Obtains written assurance, at least every 2 years, that the farm acknowledges that its food is subject to section 402 of the Federal Food, Drug, and Cosmetic Act (or, when applicable, that its food is subject to relevant laws and regulations of a country whose food safety system FDA has officially recognized as comparable or has determined to be equivalent to that of the United States).

e) If a supplier is a shell egg producer that is not subject to the requirements of part 118 of this chapter because it has less than 3,000 laying hens, the receiving facility does not need to comply with paragraphs (a) and (b) of this section if the receiving facility:
   1) Obtains written assurance that the shell eggs produced by the supplier are not subject to part 118 because the shell egg producer has less than 3,000 laying hens:
      i) Before first approving the supplier for an applicable calendar year; and
      ii) On an annual basis thereafter, by December 31 of each calendar year, for the following calendar year; and

2) Obtains written assurance, at least every 2 years, that the shell egg producer acknowledges that its food is subject to section 402 of the Federal Food, Drug, and Cosmetic Act (or, when applicable, that its food is subject to relevant laws and regulations of a country whose food safety system FDA has officially recognized as comparable or has determined to be equivalent to that of the United States).

f) There must not be any financial conflicts of interests that influence the results of the verification activities listed in § 117.410(b) and payment must not be related to the results of the activity.

§ 117.435 Onsite audit.

a) An onsite audit of a supplier must be performed by a qualified auditor.
b) If the raw material or other ingredient at the supplier is subject to one or more FDA food safety regulations, an onsite audit must consider such regulations and include a review of the supplier's written plan (*e.g.*, Hazard Analysis and Critical Control Point (HACCP) plan or other food safety plan), if any, and its implementation, for the hazard being controlled (or, when applicable, an onsite audit may consider relevant laws and regulations of a country whose food safety system FDA has officially recognized as comparable or has determined to be equivalent to that of the United States).
c) 1) The following may be substituted for an onsite audit, provided that the inspection was conducted within 1 year of the date that the onsite audit would have been required to be conducted:
   i) The written results of an appropriate inspection of the supplier for compliance with applicable FDA food safety regulations by FDA, by representatives of other Federal Agencies (such as the United States Department of Agriculture), or by representatives of State, local, tribal, or territorial agencies; or
   ii) For a foreign supplier, the written results of an inspection by FDA or the food safety authority of a country whose food safety system FDA has officially recognized as comparable or has determined to be equivalent to that of the United States.
   2) For inspections conducted by the food safety authority of a country whose food safety system FDA has officially recognized as comparable or determined to be equivalent, the food that is the subject of the onsite audit must be within the scope of the official recognition or equivalence determination, and the foreign supplier must be in, and under the regulatory oversight of, such country.
d) If the onsite audit is solely conducted to meet the requirements of this subpart by an audit agent of a certification body that is accredited in accordance with regulations in part 1, subpart M of this chapter, the audit is not subject to the requirements in those regulations.

§ 117.475 Records documenting the supply-chain program.

a) The records documenting the supply-chain program are subject to the requirements of subpart F of this part.
b) The receiving facility must review the records listed in paragraph (c) of this section in accordance with § 117.165(a)(4).
c) The receiving facility must document the following in records as applicable to its supply-chain program:
   1) The written supply-chain program;
   2) Documentation that a receiving facility that is an importer is in compliance with the foreign supplier verification program requirements under part 1, subpart L of this chapter, including documentation of verification activities conducted under § 1.506(e) of this chapter;
   3) Documentation of the approval of a supplier;
   4) Written procedures for receiving raw materials and other ingredients;
   5) Documentation demonstrating use of the written procedures for receiving raw materials and other ingredients;
   6) Documentation of the determination of the appropriate supplier verification activities for raw materials and other ingredients;
   7) Documentation of the conduct of an onsite audit. This documentation must include:
      i) The name of the supplier subject to the onsite audit;
      ii) Documentation of audit procedures;
      iii) The dates the audit was conducted;
      iv) The conclusions of the audit;
      v) Corrective actions taken in response to significant deficiencies identified during the audit; and
      vi) Documentation that the audit was conducted by a qualified auditor;
   8) Documentation of sampling and testing conducted as a supplier verification activity. This documentation must include:
      i) Identification of the raw material or other ingredient tested (including lot number, as appropriate) and the number of samples tested;
      ii) Identification of the test(s) conducted, including the analytical method(s) used;
      iii) The date(s) on which the test(s) were conducted and the date of the report;
      iv) The results of the testing;
      v) Corrective actions taken in response to detection of hazards; and
      vi) Information identifying the laboratory conducting the testing;
   9) Documentation of the review of the supplier's relevant food safety records. This documentation must include:
      i) The name of the supplier whose records were reviewed;
      ii) The date(s) of review;
      iii) The general nature of the records reviewed;

iv) The conclusions of the review; and
v) Corrective actions taken in response to significant deficiencies identified during the review;

10) Documentation of other appropriate supplier verification activities based on the supplier performance and the risk associated with the raw material or other ingredient;
11) Documentation of any determination that verification activities other than an onsite audit, and/or less frequent onsite auditing of a supplier, provide adequate assurance that the hazards are controlled when a hazard in a raw material or other ingredient will be controlled by the supplier and is one for which there is a reasonable probability that exposure to the hazard will result in serious adverse health consequences or death to humans;
12) The following documentation of an alternative verification activity for a supplier that is a qualified facility:
    i) The written assurance that the supplier is a qualified facility as defined by § 117.3, before approving the supplier and on an annual basis thereafter; and
    ii) The written assurance that the supplier is producing the raw material or other ingredient in compliance with applicable FDA food safety regulations (or, when applicable, relevant laws and regulations of a country whose food safety system FDA has officially recognized as comparable or has determined to be equivalent to that of the United States);
13) The following documentation of an alternative verification activity for a supplier that is a farm that supplies a raw material or other ingredient and is not a covered farm under part 112 of this chapter:
    i) The written assurance that supplier is not a covered farm under part 112 of this chapter in accordance with § 112.4(a), or in accordance with §§ 112.4(b) and 112.5, before approving the supplier and on an annual basis thereafter; and
    ii) The written assurance that the farm acknowledges that its food is subject to section 402 of the Federal Food, Drug, and Cosmetic Act (or, when applicable, that its food is subject to relevant laws and regulations of a country whose food safety system FDA has officially recognized as comparable or has determined to be equivalent to that of the United States);
14) The following documentation of an alternative verification activity for a supplier that is a shell egg producer that is not subject to the requirements established in part 118 of this chapter because it has less than 3,000 laying hens:
    i) The written assurance that the shell eggs provided by the supplier are not subject to part 118 of this chapter because the supplier has

less than 3,000 laying hens, before approving the supplier and on an annual basis thereafter; and

ii) The written assurance that the shell egg producer acknowledges that its food is subject to section 402 of the Federal Food, Drug, and Cosmetic Act (or, when applicable, that its food is subject to relevant laws and regulations of a country whose safety system FDA has officially recognized as comparable or has determined to be equivalent to that of the United States);

15) The written results of an appropriate inspection of the supplier for compliance with applicable FDA food safety regulations by FDA, by representatives of other Federal Agencies (such as the United States Department of Agriculture), or by representatives from State, local, tribal, or territorial agencies, or the food safety authority of another country when the results of such an inspection is substituted for an onsite audit;

16) Documentation of actions taken with respect to supplier nonconformance;

17) Documentation of verification of a supply-chain-applied control applied by an entity other than the receiving facility's supplier; and

18) When applicable, documentation of the receiving facility's review and assessment of:

   i) Applicable documentation from an entity other than the receiving facility that written procedures for receiving raw materials and other ingredients are being followed;

   ii) Applicable documentation, from an entity other than the receiving facility, of the determination of the appropriate supplier verification activities for raw materials and other ingredients;

   iii) Applicable documentation, from an entity other than the receiving facility, of conducting the appropriate supplier verification activities for raw materials and other ingredients;

   iv) Applicable documentation, from its supplier, of:

      A) The results of sampling and testing conducted by the supplier; or

      B) The results of an audit conducted by a third-party qualified auditor in accordance with §§ 117.430(f) and 117.435; and

   v) Applicable documentation, from an entity other than the receiving facility, of verification activities when a supply-chain-applied control is applied by an entity other than the receiving facility's supplier.

* * * * *

Appendix A

Food Safety Plan Checklist

The following Food Safety Plan Checklist is an updated adaptation of a Checklist submitted by the Grocery Manufacturers Association to the U.S. Government as part of a public comment about FSMA (August 29, 2011, FDA-2011-0238-0039, Appendix B, available at www.regulations.gov).

Food Safety Plan Checklist

Introduction

Under the Food Safety Modernization Act (FSMA) for human foods, registered food facilities are required to:

- Identify and evaluate the hazards that could affect the food a facility manufactures, processes, packs, or holds.
- Identify and implement preventive controls to significantly minimize or prevent the occurrence of those hazards, ensure that food is not adulterated, and prevent undeclared allergens. Preventive controls include risk-based and scientifically appropriate procedures, practices, and processes, such as Process Controls, Sanitation Preventive Controls, Food Allergen Preventive Controls, Supplier Controls, and Other Preventive Controls. A Recall Plan is also required as a control measure.
- Monitor the performance of preventive controls.
- Establish corrective actions as necessary.
- Verify that preventive controls are working.
- Maintain appropriate documents and records.
- Make appropriate documents available to U.S Food and Drug Administration (FDA) during an inspection. Facilities must have a written hazard analysis and Food Safety Plan for preventive controls. FDA has authority to inspect a facility's written records.

FSMA and Food Safety Systems: Understanding and Implementing the Rules,
First Edition. Jeffrey T. Barach.

Purpose

This checklist is provided as an aid to companies that are developing a new Food Safety Plan or revising their existing HACCP plan to be compliant with the requirements in FSMA and the regulations and guidance developed from that law. This document is not a comprehensive document on "how to" develop a Food Safety Plan nor a summary of legal requirements, but rather is a tool to assist in the many activities associated with plan development. The intent of this document is to outline major activities that should take place. Depending on the company, the activities outlined below may be undertaken by a corporate group as well as by the food safety personnel in a particular facility.

This document does not outline the only approach to considering the requirements of developing a Food Safety Plan; companies use different approaches to meet regulatory requirements.

Date:______ Reviewed By:___________ Facility: ________________

	Completed		
Item/ Activity – Principle	**YES**	**NO**	**Date**
1. Preliminary Tasks – ***Inventory and assess current operations against FSMA requirements***			
1.1 Develop a Food Safety Plan team that includes a preventive control qualified individual (PCQI)			
1.2 Describe product and intended use and/or consumer			
1.3 Describe plant operational practices – prerequisite type programs			
1.4 Develop flow diagram including receipt of raw materials, process steps, processing equipment, packaging, storage, and shipping			
1.5 Identify existing regulatory requirements being addressed (FSMA, cGMPs, Juice HACCP, Seafood HACCP, LACF, Allergen Labeling, etc.)			
1.6 Review and update all SOPs related to cGMPs			
1.7 Review and update the facilities Recall Plan			

Item/ Activity – Principle	Completed		Date
	YES	NO	
2. Hazard Analysis and Preventive Controls – ***Identify and evaluate potential hazards that require a preventive control(s)***			
2.1 Conduct a hazard analysis			
2.2 Determine preventive controls, which may include (but are not limited to) the following: Process Controls – Critical Control Points (CCPs) Sanitation Preventive Controls Food Allergen Preventive Controls Supplier Preventive Controls Other Preventive Controls			
2.3 Consider design modifications that could assist in managing/mitigating any potential hazards			
3. Monitoring – ***Establish monitoring practices for each preventive control***			
3.1 Identify monitoring activities associated with preventive controls, as appropriate, to assure that identified hazards will be eliminated, significantly minimized, or prevented			
3.2 Define and determine critical limits for CCPs, if any			
3.3 Establish critical parameters for all preventive controls			
4. Corrective Actions – ***Establish procedures for corrective actions to be taken when preventive controls are not properly implemented or are found to be ineffective***			
4.1 Identify procedures for corrective actions to be taken when preventive controls are not properly implemented or are found to be ineffective			

Item/ Activity – Principle	Completed		Date
	YES	NO	
4.2 Establish procedures for documenting corrective actions (e.g., a corrective action form)			
4.3 Establish procedures to ensure appropriate actions are taken to reduce the likelihood of a recurrence of the deviation (focus on means to find the root cause of the problem)			
4.4 Establish procedures to evaluate all affected food for safety and prevent it from entering commerce if its safety cannot be established			
4.5 Establish a Recall Plan, if none is available			
5. Verification and Validation ***Establish procedures to verify that the preventive controls are effective and that the Food Safety Plan is working correctly***			
5.1 Establish the scientific or other basis, as appropriate, that documents the validity of the preventive control measure(s) and that hazards are adequately prevented, eliminated, or reduced to a level that ensures food safety			
5.2 Implement an initial audit that verifies the Food Safety Plan is designed properly to control hazards			
5.3 Establish documented, periodic reanalysis of the plan to ensure it is still relevant when (1) significant changes that create a foreseeable potential to affect food safety occur, or (2) every 3 years, whichever is earlier			
5.4 As appropriate, establish environmental monitoring and product testing programs as verification activities			
6. Records – ***Establish effective record-keeping procedures that document the Food Safety Plan***			

Item/ Activity – Principle	Completed		Date
	YES	NO	
6.1 Identify record-keeping requirements from regulations (e.g., FSMA, BT Act, HACCP, LACF)			
6.2 Develop means to establish and maintain essential Food Safety Plan records, and decide where records will be kept (e.g., in a Corporate location or at a particular facility) and how long			
7. **Training –** ***Establish effective training programs for management and line-workers***			
7.1 Ensure management communicates a "Food Safety Culture" to all employees			
7.2 Provide training for management and staff who will design and oversee the Food Safety Plan			
7.3 Establish a line-worker training program for operators that will carry out the Food Safety Plan			
7.4 Document training was received and assess its effectiveness			

Items Needing Follow-up

Describe Item	Describe Action to Correct	Who is Responsible for Follow-up?	When Will Correction be Made?	Completed on (Date)

References

1 Barach, J.T. (ed.), 2013, *A systems approach using preventive controls for safe food production: Part 1 establishing a food safety system, GMA*, Washington, DC.
2 FDA Food Safety Modernization Act, viewed 11 July 2016, http://www.fda.gov/Food/FoodSafety/FSMA/default.htm.
3 FDA Food Defense, viewed 11 July 2016, http://www.fda.gov/Food/FoodDefense/default.htm.
4 Food Safety Management Systems – requirements for any organization in the food chain, ISO 22000:2005-09-01.
5 Joint FAO/WHO Codex Alimentarius Commission World Health Organization Food and Agriculture Organization of the United Nations, 2004, Annex: Hazard analysis and critical control point (HACCP) system and guidelines for its application, 2003. 3rd edn., in *Codex alimentarius food hygiene basic texts FAO/WHO 2004*, Rome.
6 Recommended international code of practice general principles of food hygiene – CAC/RCP 1-1969, Rev. 4-2003.

Appendix B

HACCP Principles and Application Guidelines

National Advisory Committee on Microbiological Criteria for Foods

The National Advisory Committee on Microbiological Criteria for Foods (NACMCF) is an advisory committee chartered under the U.S. Department of Agriculture (USDA) and comprised of participants from the USDA (Food Safety and Inspection Service), Department of Health and Human Services (U.S. Food and Drug Administration and the Centers for Disease Control and Prevention), the Department of Commerce (National Marine Fisheries Service), the Department of Defense (Office of the Army Surgeon General), academia, industry and state employees. NACMCF provides guidance and recommendations to the Secretary of Agriculture and the Secretary of Health and Human Services regarding the microbiological safety of foods.

Executive Summary

The National Advisory Committee on Microbiological Criteria for Foods (Committee) reconvened a Hazard Analysis and Critical Control Point (HACCP) Working Group in 1995. The primary goal was to review the Committee's November 1992 HACCP document, comparing it to current HACCP guidance prepared by the Codex Committee on Food Hygiene. Based upon its review, the Committee made the HACCP principles more concise; revised and added definitions; included sections on prerequisite programs, education and training, and implementation and maintenance of the HACCP plan; revised and provided a more detailed explanation of the application of HACCP principles; and provided an additional decision tree for identifying critical control points (CCPs).

The Committee again endorses HACCP as an effective and rational means of assuring food safety from harvest to consumption. Preventing problems from occurring is the paramount goal underlying any HACCP system. Seven basic principles are employed in the development of HACCP plans that meet the

FSMA and Food Safety Systems: Understanding and Implementing the Rules,
First Edition. Jeffrey T. Barach.

stated goal. These principles include hazard analysis, CCP identification, establishing critical limits, monitoring procedures, corrective actions, verification procedures, and record-keeping and documentation. Under such systems, if a deviation occurs indicating that control has been lost, the deviation is detected and appropriate steps are taken to reestablish control in a timely manner to assure that potentially hazardous products do not reach the consumer.

In the application of HACCP, the use of microbiological testing is seldom an effective means of monitoring CCPs because of the time required to obtain results. In most instances, monitoring of CCPs can best be accomplished through the use of physical and chemical tests, and through visual observations. Microbiological criteria do, however, play a role in verifying that the overall HACCP system is working.

The Committee believes that the HACCP principles should be standardized to provide uniformity in training and applying the HACCP system by industry and government. In accordance with the National Academy of Sciences recommendation, the HACCP system must be developed by each food establishment and tailored to its individual product, processing and distribution conditions.

In keeping with the Committee's charge to provide recommendations to its sponsoring agencies regarding microbiological food safety issues, this document focuses on this area. The Committee recognizes that in order to assure food safety, properly designed HACCP systems must also consider chemical and physical hazards in addition to other biological hazards.

For a successful HACCP program to be properly implemented, management must be committed to a HACCP approach. A commitment by management will indicate an awareness of the benefits and costs of HACCP and include education and training of employees. Benefits, in addition to enhanced assurance of food safety, are better use of resources and timely response to problems.

The Committee designed this document to guide the food industry and advise its sponsoring agencies in the implementation of HACCP systems. See document at: http://www.fda.gov/Food/GuidanceRegulation/HACCP/ucm2006801.htm

Definitions

CCP Decision Tree:
A sequence of questions to assist in determining whether a control point is a CCP.

Control:

a) To manage the conditions of an operation to maintain compliance with established criteria.

b) The state where correct procedures are being followed and criteria are being met.

Control Measure:
Any action or activity that can be used to prevent, eliminate or reduce a significant hazard.

Control Point:
Any step at which biological, chemical, or physical factors can be controlled.

Corrective Action:
Procedures followed when a deviation occurs.

Criterion:
A requirement on which a judgement or decision can be based.

Critical Control Point:
A step at which control can be applied and is essential to prevent or eliminate a food safety hazard or reduce it to an acceptable level.

Critical Limit:
A maximum and/or minimum value to which a biological, chemical or physical parameter must be controlled at a CCP to prevent, eliminate or reduce to an acceptable level the occurrence of a food safety hazard.

Deviation:
Failure to meet a critical limit.

HACCP:
A systematic approach to the identification, evaluation, and control of food safety hazards.

HACCP Plan:
The written document which is based upon the principles of HACCP and which delineates the procedures to be followed.

HACCP System:
The result of the implementation of the HACCP Plan.

HACCP Team:
The group of people who are responsible for developing, implementing and maintaining the HACCP system.

Hazard:
A biological, chemical, or physical agent that is reasonably likely to cause illness or injury in the absence of its control.

Hazard Analysis:
The process of collecting and evaluating information on hazards associated with the food under consideration to decide which are significant and must be addressed in the HACCP plan.

Monitor:
To conduct a planned sequence of observations or measurements to assess whether a CCP is under control and to produce an accurate record for future use in verification.

Prerequisite Programs:
Procedures, including Good Manufacturing Practices, that address operational conditions providing the foundation for the HACCP system.

Severity:
The seriousness of the effect(s) of a hazard.

Step:
A point, procedure, operation or stage in the food system from primary production to final consumption.

Validation:
That element of verification focused on collecting and evaluating scientific and technical information to determine if the HACCP plan, when properly implemented, will effectively control the hazards.

Verification:
Those activities, other than monitoring, that determine the validity of the HACCP plan and that the system is operating according to the plan.

HACCP Principles

HACCP is a systematic approach to the identification, evaluation, and control of food safety hazards based on the following seven principles:

Principle 1: Conduct a hazard analysis.
Principle 2: Determine the critical control points (CCPs).
Principle 3: Establish critical limits.
Principle 4: Establish monitoring procedures.
Principle 5: Establish corrective actions.

Principle 6: Establish verification procedures.
Principle 7: Establish record-keeping and documentation procedures.

Guidelines for Application of HACCP Principles

Introduction

HACCP is a management system in which food safety is addressed through the analysis and control of biological, chemical, and physical hazards from raw material production, procurement and handling, to manufacturing, distribution and consumption of the finished product. For successful implementation of a HACCP plan, management must be strongly committed to the HACCP concept. A firm commitment to HACCP by top management provides company employees with a sense of the importance of producing safe food.

HACCP is designed for use in all segments of the food industry from growing, harvesting, processing, manufacturing, distributing, and merchandising to preparing food for consumption. Prerequisite programs such as current Good Manufacturing Practices (cGMPs) are an essential foundation for the development and implementation of successful HACCP plans. Food safety systems based on the HACCP principles have been successfully applied in food processing plants, retail food stores, and food service operations. The seven principles of HACCP have been universally accepted by government agencies, trade associations and the food industry around the world.

The following guidelines will facilitate the development and implementation of effective HACCP plans. While the specific application of HACCP to manufacturing facilities is emphasized here, these guidelines should be applied as appropriate to each segment of the food industry under consideration.

Prerequisite Programs

The production of safe food products requires that the HACCP system be built upon a solid foundation of prerequisite programs. Examples of common prerequisite programs are listed in Appendix A. Each segment of the food industry must provide the conditions necessary to protect food while it is under their control. This has traditionally been accomplished through the application of cGMPs. These conditions and practices are now considered to be prerequisite to the development and implementation of effective HACCP plans. Prerequisite programs provide the basic environmental and operating conditions that are necessary for the production of safe, wholesome food. Many of the conditions and practices are specified in federal, state and local regulations and guidelines (e.g., cGMPs and Food Code). The Codex Alimentarius General Principles of Food Hygiene describe the basic conditions and practices expected for foods intended for international trade. In addition to the requirements specified in regulations, industry often adopts policies and procedures that are specific to their operations. Many of these are proprietary. While prerequisite programs may impact upon the safety of a food, they also are concerned with ensuring that foods are wholesome and suitable for consumption (Appendix A). HACCP plans are narrower in scope, being limited to ensuring food is safe to consume.

The existence and effectiveness of prerequisite programs should be assessed during the design and implementation of each HACCP plan. All prerequisite programs should be documented and regularly audited. Prerequisite programs are established and managed separately from the HACCP plan. Certain aspects, however, of a prerequisite program may be incorporated into a HACCP plan. For example, many establishments have preventive maintenance procedures for processing equipment to avoid unexpected equipment failure and loss of production. During the development of a HACCP plan, the HACCP team may decide that the routine maintenance and calibration of an oven should be included in the plan as an activity of verification. This would further ensure that all the food in the oven is cooked to the minimum internal temperature that is necessary for food safety.

Education and Training

The success of a HACCP system depends on educating and training management and employees in the importance of their role in producing safe foods. This should also include information [about] the control of foodborne hazards related to all stages of the food chain. It is important to recognize that employees must first understand what HACCP is and then learn the skills necessary to make it function properly. Specific training activities should include working instructions and procedures that outline the tasks of employees monitoring each CCP.

Management must provide adequate time for thorough education and training. Personnel must be given the materials and equipment necessary to perform these tasks. Effective training is an important prerequisite to successful implementation of a HACCP plan.

Developing a HACCP Plan

The format of HACCP plans will vary. In many cases the plans will be product and process specific. However, some plans may use a unit operations approach. Generic HACCP plans can serve as useful guides in the development of process and product HACCP plans; however, it is essential that the unique conditions within each facility be considered during the development of all components of the HACCP plan.

In the development of a HACCP plan, five preliminary tasks need to be accomplished before the application of the HACCP principles to a specific product and process. The five preliminary tasks are given in Figure 1.

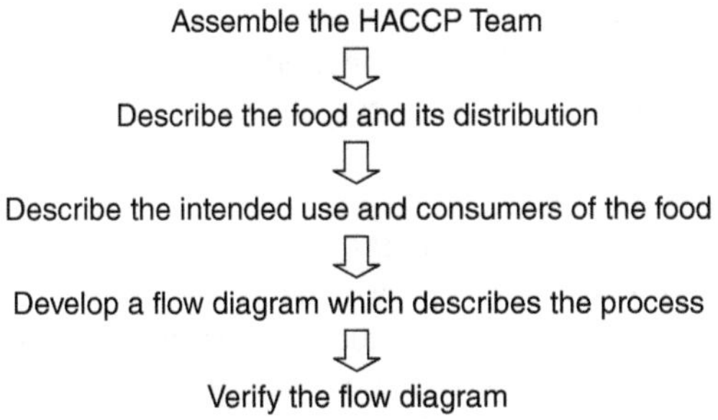

Assemble the HACCP Team

⇩

Describe the food and its distribution

⇩

Describe the intended use and consumers of the food

⇩

Develop a flow diagram which describes the process

⇩

Verify the flow diagram

Figure 1 Preliminary Tasks in the Development of the HACCP Plan.

Assemble the HACCP Team

The first task in developing a HACCP plan is to assemble a HACCP team consisting of individuals who have specific knowledge and expertise appropriate to the product and process. It is the team's responsibility to develop the HACCP plan. The team should be multi disciplinary and include individuals from areas such as engineering, production, sanitation, quality assurance, and food microbiology. The team should also include local personnel who are involved in the operation as they are more familiar with the variability and limitations of the operation. In addition, this fosters a sense of ownership among those who must implement the plan. The HACCP team may need assistance from outside experts who are knowledgeable in the potential biological, chemical and/or physical hazards associated with the product and the process. However, a plan which is developed totally by outside sources may be erroneous, incomplete, and lacking in support at the local level.

Due to the technical nature of the information required for hazard analysis, it is recommended that experts who are knowledgeable in the food process should either participate in or verify the completeness of the hazard analysis and the HACCP plan. Such individuals should have the knowledge and experience to correctly: (a) conduct a hazard analysis; (b) identify potential hazards; (c) identify hazards which must be controlled; (d) recommend controls, critical limits, and procedures for monitoring and verification; (e) recommend appropriate corrective actions when a deviation occurs; (f) recommend research related to the HACCP plan if important information is not known; and (g) validate the HACCP plan.

Describe the food and its distribution

The HACCP team first describes the food. This consists of a general description of the food, ingredients, and processing methods. The method of distribution should be described along with information on whether the food is to be distributed frozen, refrigerated, or at ambient temperature.

Describe the intended use and consumers of the food

Describe the normal expected use of the food. The intended consumers may be the general public or a particular segment of the population (e.g., infants, immunocompromised individuals, the elderly, etc.).

Develop a flow diagram which describes the process

The purpose of a flow diagram is to provide a clear, simple outline of the steps involved in the process. The scope of the flow diagram must cover all the steps in the process which are directly under the control of the establishment. In addition, the flow diagram can include steps in the food chain which are before and after the processing that occurs in the establishment. The flow diagram need not be as complex as engineering drawings. A block type flow diagram is sufficiently descriptive (see Appendix B). Also, a simple schematic of the facility is often useful in understanding and evaluating product and process flow.

Verify the flow diagram

The HACCP team should perform an on-site review of the operation to verify the accuracy and completeness of the flow diagram. Modifications should be made to the flow diagram as necessary and documented.

After these five preliminary tasks have been completed, the seven principles of HACCP are applied.

Conduct a hazard analysis (Principle 1)

After addressing the preliminary tasks discussed above, the HACCP team conducts a hazard analysis and identifies appropriate control measures. The purpose of the hazard analysis is to develop a list of hazards which are of such significance that they are reasonably likely to cause injury or illness if not effectively controlled. Hazards that are not reasonably likely to occur would not require further consideration within a HACCP plan. It is important to consider in the hazard analysis the ingredients and raw materials, each step in the process, product storage and distribution, and final preparation and use by the consumer. When conducting a hazard analysis, safety concerns must be differentiated from quality concerns. A hazard is defined as a biological, chemical or physical agent that is reasonably likely to cause illness or injury in the absence of its control. Thus, the word hazard as used in this document is limited to safety.

A thorough hazard analysis is the key to preparing an effective HACCP plan. If the hazard analysis is not done correctly and the hazards warranting control within the HACCP system are not identified, the plan will not be effective regardless of how well it is followed.

The hazard analysis and identification of associated control measures accomplish three objectives: Those hazards and associated control measures are identified. The analysis may identify needed modifications to a process or product so that product safety is further assured or improved. The analysis provides a basis for determining CCPs in Principle 2.

The process of conducting a hazard analysis involves two stages. The first, hazard identification, can be regarded as a brain storming session. During this stage, the HACCP team reviews the ingredients used in the product, the activities conducted at each step in the process and the equipment used, the final product and its method of storage and distribution, and the intended use and consumers of the product. Based on this review, the team develops a list of potential biological, chemical or physical hazards which may be introduced, increased, or controlled at each step in the production process. Appendix C lists examples of questions that may be helpful to consider when identifying potential hazards. Hazard identification focuses on developing a list of potential hazards associated with each process step under direct control of the food operation. A knowledge of any adverse health-related events historically associated with the product will be of value in this exercise.

After the list of potential hazards is assembled, stage two, the hazard evaluation, is conducted. In stage two of the hazard analysis, the HACCP team decides which potential hazards must be addressed in the HACCP plan. During this stage, each potential hazard is evaluated based on the severity of the potential hazard and its likely occurrence. Severity is the seriousness of the consequences of exposure to the hazard. Considerations of severity (e.g., impact of sequelae, and magnitude and duration of illness or injury) can be helpful in understanding the public health impact of the hazard. Consideration of the likely occurrence is usually based upon a combination of experience, epidemiological data, and information in the technical literature. When conducting the hazard evaluation, it is helpful to consider the likelihood of exposure and severity of the potential consequences if the hazard is not properly controlled. In addition, consideration should be given to the effects of short term as well as long term exposure to the potential hazard. Such considerations do not include common dietary choices which lie outside of HACCP. During the evaluation of each potential hazard, the food, its method of preparation, transportation, storage and persons likely to consume the product should be considered to determine how each of these factors may influence the likely occurrence and severity of the hazard being controlled. The team must consider the influence of likely procedures for food preparation and storage and whether the intended consumers are susceptible to a potential hazard. However, there may be differences of opinion, even among experts, as to the likely occurrence and severity of a hazard. The HACCP team may have to rely upon the opinion of experts who assist in the development of the HACCP plan.

Hazards identified in one operation or facility may not be significant in another operation producing the same or a similar product. For example, due to differences in equipment and/or an effective maintenance program, the probability of metal contamination may be significant in one facility but not in another. A summary of the HACCP team deliberations and the rationale developed during the hazard analysis should be kept for future reference. This information will be useful during future reviews and updates of the hazard analysis and the HACCP plan.

Appendix D gives three examples of using a logic sequence in conducting a hazard analysis. While these examples relate to biological hazards, chemical and physical hazards are equally important to consider. Appendix D is for illustration purposes to further explain the stages of hazard analysis for identifying hazards. Hazard identification and evaluation as outlined in Appendix D may eventually be assisted by biological risk assessments as they become available. While the process and output of a risk assessment (NACMCF, 1997)(1) is significantly different from a hazard analysis, the identification of hazards of concern and the hazard evaluation may be facilitated by information from risk assessments. Thus, as risk assessments addressing specific hazards or control factors become available, the HACCP team should take these into consideration.

Upon completion of the hazard analysis, the hazards associated with each step in the production of the food should be listed along with any measure(s) that are used to control the hazard(s). The term control measure is used because not all hazards can be prevented, but virtually all can be controlled. More than one control measure may be required for a specific hazard. On the other hand, more than one hazard may be addressed by a specific control measure (e.g. pasteurization of milk).

For example, if a HACCP team were to conduct a hazard analysis for the production of frozen cooked beef patties (Appendices B and D), enteric pathogens (e.g., *Salmonella* and verotoxin-producing *Escherichia coli*) in the raw meat would be identified as hazards. Cooking is a control measure which can be used to eliminate these hazards. The following (Table 1.) is an excerpt from a hazard analysis summary table for this product.

Table 1 Excerpt From a Hazard Analysis Summary Table.

Step	Potential Hazard(s)	Justification	Hazard to be Addressed in Plan? (Y/N)	Control Measure(s)
5. Cooking	Enteric pathogens: e.g., *Salmonella*, verotoxigenic-*E. coli*	Enteric pathogens have been associated with outbreaks of foodborne illness from undercooked ground beef	Y	Cooking

The hazard analysis summary could be presented in several different ways. One format is a table such as the one given above. Another could be a narrative summary of the HACCP team's hazard analysis considerations and a summary table listing only the hazards and associated control measures.

Determine critical control points (CCPs) (Principle 2)

A critical control point is defined as a step at which control can be applied and is essential to prevent or eliminate a food safety hazard or reduce it to an acceptable level. The potential hazards that are reasonably likely to cause illness or injury in the absence of their control must be addressed in determining CCPs.

Complete and accurate identification of CCPs is fundamental to controlling food safety hazards. The information developed during the hazard analysis is essential for the HACCP team in identifying which steps in the process are CCPs. One strategy to facilitate the identification of each CCP is the use of a CCP decision tree (Examples of decision trees are given in Appendices E and F). Although application of the CCP decision tree can be useful in determining if a particular step is a CCP for a previously identified hazard, it is merely a tool and not a mandatory element of HACCP. A CCP decision tree is not a substitute for expert knowledge.

Critical control points are located at any step where hazards can be either prevented, eliminated, or reduced to acceptable levels. Examples of CCPs may include: thermal processing, chilling, testing ingredients for chemical residues, product formulation control, and testing product for metal contaminants. CCPs must be carefully developed and documented. In addition, they must be used only for purposes of product safety. For example, a specified heat process, at a given time and temperature designed to destroy a specific microbiological pathogen, could be a CCP. Likewise, refrigeration of a precooked food to prevent hazardous microorganisms from multiplying, or the adjustment of a food to a pH necessary to prevent toxin formation could also be CCPs. Different facilities preparing similar food items can differ in the hazards identified and the steps which are CCPs. This can be due to differences in each facility's layout, equipment, selection of ingredients, processes employed, etc.

Establish critical limits (Principle 3)

A critical limit is a maximum and/or minimum value to which a biological, chemical or physical parameter must be controlled at a CCP to prevent, eliminate or reduce to an acceptable level the occurrence of a food safety hazard. A critical limit is used to distinguish between safe and unsafe operating conditions at a CCP. Critical limits should not be confused with operational limits which are established for reasons other than food safety.

Each CCP will have one or more control measures to assure that the identified hazards are prevented, eliminated or reduced to acceptable levels. Each control measure has one or more associated critical limits. Critical limits may be based upon factors such as: temperature, time, physical dimensions, humidity, moisture level, water activity (a_w), pH, titratable acidity, salt concentration, available chlorine, viscosity, preservatives, or sensory information such as aroma and visual appearance. Critical limits must be scientifically based. For each CCP, there is at least one criterion for food safety that is to be met. An example of a criterion is a specific lethality of a cooking process such as a 5D reduction in *Salmonella*. The critical limits and criteria for food safety may be derived from sources such as regulatory standards and guidelines, literature surveys, experimental results, and experts.

An example is the cooking of beef patties (Appendix B). The process should be designed to ensure the production of a safe product. The hazard analysis for cooked meat patties identified enteric pathogens (e.g., verotoxigenic *E. coli* such as *E. coli* O157:H7, and salmonellae) as significant biological hazards. Furthermore, cooking is the step in the process at which control can be applied to reduce the enteric pathogens to an acceptable level. To ensure that an acceptable level is consistently achieved, accurate information is needed on the probable number of the pathogens in the raw patties, their heat resistance, the factors that influence the heating of the patties, and the area of the patty which heats the slowest. Collectively, this information forms the scientific basis for the critical limits that are established. Some of the factors that may affect the thermal destruction of enteric pathogens are listed in the following table (Table 2.). In this example, the HACCP team concluded that a thermal process equivalent to 155° F for 16 seconds would be necessary to assure the safety of this product. To ensure that this time and temperature are attained, the HACCP team for one facility determined that it would be necessary to establish critical limits for the oven temperature and humidity, belt speed (time in oven), patty thickness and composition (e.g., all beef, beef and other ingredients). Control of these factors enables the facility to produce a wide variety of cooked patties, all of which will be processed to a minimum internal temperature of 155° F for 16 seconds. In another facility, the HACCP team may conclude that the best approach is to use the internal patty temperature of 155° F and hold for 16 seconds as critical limits. In this second facility the internal temperature and

Table 2 Excerpt from a HACCP Plan.

Process Step	CCP	Critical Limits
5. Cooking	YES	Oven temperature:___° F Time; rate of heating and cooling (belt speed in ft/min):____ft/min Patty thickness:____in. Patty composition: e.g. all beef Oven humidity:____% RH

hold time of the patties are monitored at a frequency to ensure that the critical limits are constantly met as they exit the oven. The example given below applies to the first facility.

Establish monitoring procedures (Principle 4)

Monitoring is a planned sequence of observations or measurements to assess whether a CCP is under control and to produce an accurate record for future use in verification. Monitoring serves three main purposes. First, monitoring is essential to food safety management in that it facilitates tracking of the operation. If monitoring indicates that there is a trend towards loss of control, then action can be taken to bring the process back into control before a deviation from a critical limit occurs. Second, monitoring is used to determine when there is loss of control and a deviation occurs at a CCP, i.e., exceeding or not meeting a critical limit. When a deviation occurs, an appropriate corrective action must be taken. Third, it provides written documentation for use in verification.

An unsafe food may result if a process is not properly controlled and a deviation occurs. Because of the potentially serious consequences of a critical limit deviation, monitoring procedures must be effective. Ideally, monitoring should be continuous, which is possible with many types of physical and chemical methods. For example, the temperature and time for the scheduled thermal process of low-acid canned foods is recorded continuously on temperature recording charts. If the temperature falls below the scheduled temperature or the time is insufficient, as recorded on the chart, the product from the retort is retained and the disposition determined as in Principle 5. Likewise, pH measurement may be performed continually in fluids or by testing each batch before processing. There are many ways to monitor critical limits on a continuous or batch basis and record the data on charts. Continuous monitoring is always preferred when feasible. Monitoring equipment must be carefully calibrated for accuracy.

Assignment of the responsibility for monitoring is an important consideration for each CCP. Specific assignments will depend on the number of CCPs and control measures and the complexity of monitoring. Personnel who monitor CCPs are often associated with production (e.g., line supervisors, selected line workers and maintenance personnel) and, as required, quality control personnel. Those individuals must be trained in the monitoring technique for which they are responsible, fully understand the purpose and importance of monitoring, be unbiased in monitoring and reporting, and accurately report the results of monitoring. In addition, employees should be trained in procedures to follow when there is a trend towards loss of control so that adjustments can be made in a timely manner to assure that the process remains under control. The person responsible for monitoring must also immediately report a process or product that does not meet critical limits.

All records and documents associated with CCP monitoring should be dated and signed or initialed by the person doing the monitoring.

When it is not possible to monitor a CCP on a continuous basis, it is necessary to establish a monitoring frequency and procedure that will be reliable enough to indicate that the CCP is under control. Statistically designed data collection or sampling systems lend themselves to this purpose.

Most monitoring procedures need to be rapid because they relate to on-line, "real-time" processes and there will not be time for lengthy analytical testing. Examples of monitoring activities include: visual observations and measurement of temperature, time, pH, and moisture level.

Microbiological tests are seldom effective for monitoring due to their time-consuming nature and problems with assuring detection of contaminants. Physical and chemical measurements are often preferred because they are rapid and usually more effective for assuring control of microbiological hazards. For example, the safety of pasteurized milk is based upon measurements of time and temperature of heating rather than testing the heated milk to assure the absence of surviving pathogens.

With certain foods, processes, ingredients, or imports, there may be no alternative to microbiological testing. However, it is important to recognize that a sampling protocol that is adequate to reliably detect low levels of pathogens is seldom possible because of the large number of samples needed. This sampling limitation could result in a false sense of security by those who use an inadequate sampling protocol. In addition, there are technical limitations in many laboratory procedures for detecting and quantitating pathogens and/or their toxins.

Establish corrective actions (Principle 5)

The HACCP system for food safety management is designed to identify health hazards and to establish strategies to prevent, eliminate, or reduce their occurrence. However, ideal circumstances do not always prevail and deviations from established processes may occur. An important purpose of corrective actions is to prevent foods which may be hazardous from reaching consumers. Where there is a deviation from established critical limits, corrective actions are necessary. Therefore, corrective actions should include the following elements: (a) determine and correct the cause of non-compliance; (b) determine the disposition of non-compliant product and (c) record the corrective actions that have been taken. Specific corrective actions should be developed in advance for each CCP and included in the HACCP plan. As a minimum, the HACCP plan should specify what is done when a deviation occurs, who is responsible for implementing the corrective actions, and that a record will be developed and maintained of the actions taken. Individuals who have a thorough understanding of the process, product and HACCP plan should be assigned the responsibility for oversight of corrective actions. As appropriate, experts may

Activity	Frequency	Responsibility	Reviewer
Verification Activities Scheduling	Yearly or Upon HACCP System Change	HACCP Coordinator	Plant Manager
Initial Validation of HACCP Plan	Prior to and During Initial Implementation of Plan	Independent Expert(s)[(a)]	HACCP Team
Subsequent validation of HACCP Plan	When Critical Limits Changed, Significant Changes in Process, Equipment Changed, After System Failure, etc.	Independent Expert(s)[(a)]	HACCP Team
Verification of CCP Monitoring as Described in the Plan (e.g., monitoring of patty cooking temperature)	According to HACCP Plan (e.g., once per shift)	According to HACCP Plan (e.g., Line Supervisor)	According to HACCP Plan (e.g., Quality Control)
Review of Monitoring, Corrective Action Records to Show Compliance with the Plan	Monthly	Quality Assurance	HACCP Team
Comprehensive HACCP System Verification	Yearly	Independent Expert(s)[(a)]	Plant Manager

[(a)] Done by others than the team writing and implementing the plan. May require additional technical expertise as well as laboratory and plant test studies.

Figure 2 Example of a Company Established HACCP Verification Schedule.

be consulted to review the information available and to assist in determining disposition of non-compliant product.

Establish verification procedures (Principle 6)

Verification is defined as those activities, other than monitoring, that determine the validity of the HACCP plan and that the system is operating according to the plan. The NAS (1985) (2) pointed out that the major infusion of science in a HACCP system centers on proper identification of the hazards, critical control points, critical limits, and instituting proper verification procedures. These processes should take place during the development and implementation of the HACCP plans and maintenance of the HACCP system. An example of a verification schedule is given in Figure 2.

One aspect of verification is evaluating whether the facility's HACCP system is functioning according to the HACCP plan. An effective HACCP system requires little end-product testing, since sufficient validated safeguards are built in early in the process. Therefore, rather than relying on end-product testing, firms should rely on frequent reviews of their HACCP plan, verification that the HACCP plan is being correctly followed, and review of CCP monitoring and corrective action records.

Another important aspect of verification is the initial validation of the HACCP plan to determine that the plan is scientifically and technically sound, that all hazards have been identified and that if the HACCP plan is properly implemented these hazards will be effectively controlled. Information needed to validate the HACCP plan often include (1) expert advice and scientific studies and (2) in-plant observations, measurements, and evaluations. For example, validation of the cooking process for beef patties should include the scientific justification of the heating times and temperatures needed to obtain an appropriate destruction of pathogenic microorganisms (i.e., enteric pathogens) and studies to confirm that the conditions of cooking will deliver the required time and temperature to each beef patty.

Subsequent validations are performed and documented by a HACCP team or an independent expert as needed. For example, validations are conducted when there is an unexplained system failure; a significant product, process or packaging change occurs; or new hazards are recognized.

In addition, a periodic comprehensive verification of the HACCP system should be conducted by an unbiased, independent authority. Such authorities can be internal or external to the food operation. This should include a technical evaluation of the hazard analysis and each element of the HACCP plan as well as on-site review of all flow diagrams and appropriate records from operation of the plan. A comprehensive verification is independent of other verification procedures and must be performed to ensure that the HACCP plan is resulting in the control of the hazards. If the results of the comprehensive verification identifies deficiencies, the HACCP team modifies the HACCP plan as necessary.

Verification activities are carried out by individuals within a company, third party experts, and regulatory agencies. It is important that individuals doing verification have appropriate technical expertise to perform this function. The role of regulatory and industry in HACCP was further described by the NACMCF (1994)[3].

Examples of verification activities are included as Appendix G.

Establish record-keeping and documentation procedures (Principle 7)

Generally, the records maintained for the HACCP System should include the following:

1) A summary of the hazard analysis, including the rationale for determining hazards and control measures.

2) The HACCP Plan
   Listing of the HACCP team and assigned responsibilities.
   Description of the food, its distribution, intended use, and consumer.
   Verified flow diagram.
   HACCP Plan Summary Table (see below) that includes information for:
   Steps in the process that are CCPs
   The hazard(s) of concern.
   Critical limits
   Monitoring*
   Corrective actions*
   Verification procedures and schedule*
   Record-keeping procedures*
   * A brief summary of position responsible for performing the activity and the procedures and frequency should be provided
3) Support documentation such as validation records.
4) Records that are generated during the operation of the plan.

The following is an Example of a HACCP Plan Summary Table

CCP	Hazards	Critical limit(s)	Monitoring	Corrective Actions	Verification	Records

Examples of HACCP records are given in Appendix H.

Implementation and Maintenance of the HACCP Plan

The successful implementation of a HACCP plan is facilitated by commitment from top management. The next step is to establish a plan that describes the individuals responsible for developing, implementing and maintaining the HACCP system. Initially, the HACCP coordinator and team are selected and trained as necessary. The team is then responsible for developing the initial plan and coordinating its implementation. Product teams can be appointed to develop HACCP plans for specific products. An important aspect in developing these teams is to assure that they have appropriate training. The workers who will be responsible for monitoring need to be adequately trained. Upon completion of the HACCP plan, operator procedures, forms and procedures for monitoring and corrective action are developed. Often it is a good idea to develop a timeline for the activities involved in the initial implementation of the HACCP plan. Implementation of the HACCP system involves the continual application of the monitoring, record-keeping, corrective action procedures and other activities as described in the HACCP plan.

Maintaining an effective HACCP system depends largely on regularly scheduled verification activities. The HACCP plan should be updated and revised as needed. An important aspect of maintaining the HACCP system is to assure that all individuals involved are properly trained so they understand their role and can effectively fulfill their responsibilities.

References

1 National Advisory Committee on Microbiological Criteria for Foods. 1997. The principles of risk assessment for illness caused by foodborne biological agents. Adopted April 4, 1997.
2 An Evaluation of the Role of Microbiological Criteria for Foods and Food Ingredients. 1985. National Academy of Sciences, National Academy Press, Washington, DC.
3 National Advisory Committee on Microbiological Criteria for Foods. 1994. The role of regulatory agencies and industry in HACCP. *Int. J. Food Microbiol.* 21:187–195.

For Appendices referenced in this publication, see:
(http://www.fda.gov/Food/GuidanceRegulation/HACCP/ucm2006801.htm)

APPENDIX A – Examples of common prerequisite programs
APPENDIX B – Example of a flow diagram for the production of frozen cooked beef patties.
APPENDIX C – Examples of questions to be considered when conducting a hazard analysis
APPENDIX D – Examples of how the stages of hazard analysis are used to identify and evaluate hazards
APPENDIX E – Example I of a CCP decision tree
APPENDIX F – Example II of a CCP decision tree
APPENDIX G – Examples of verification activities
APPENDIX H – Examples of HACCP records

Glossary

acid foods or acidified foods Foods that have an equilibrium pH of 4.6 or below.

audit The systematic, independent, and documented examination (through observation, investigation, records review, discussions with employees of the audited entity, and, as appropriate, sampling and laboratory analysis) to assess a supplier's food safety processes and procedures.

correction An action to identify and correct a problem that occurred during the production of food, without other actions associated with a corrective action procedure (such as actions to reduce the likelihood that the problem will recur, evaluate all affected food for safety, and prevent affected food from entering commerce).

corrective actions Procedures that must be taken if preventive controls are not properly implemented and deviations occur.

Critical Control Point (CCP) A point, step, or procedure in a food process at which control can be applied and is essential to prevent or eliminate a food safety hazard or reduce it to an acceptable level.

critical limit A maximum and/or minimum value to which a biological, chemical, physical, or radiological parameter must be controlled at a CCP to prevent, eliminate, or reduce to an acceptable level the occurrence of a food safety hazard.

cross-contact The unintentional incorporation of a food allergen into a food.

deviation Failure to meet a critical limit.

environmental pathogen A pathogen capable of surviving and persisting within the manufacturing, processing, packing, or holding environment such that food may be contaminated and may result in foodborne illness if that food is consumed without treatment to significantly minimize the environmental pathogen.

facility A domestic facility or a foreign facility that is required to register under section 415 of the Federal Food, Drug and Cosmetic Act, in accordance with the requirements of 21 CFR part 1, subpart H.

FSMA and Food Safety Systems: Understanding and Implementing the Rules,
First Edition. Jeffrey T. Barach.

FDA Food and Drug Administration.
food allergen A major food allergen as defined in section 201(qq) of the Federal Food, Drug and Cosmetic Act and includes raw materials and ingredients.
food-contact surfaces Those surfaces that contact human food and those surfaces from which drainage, or other transfer, onto the food or onto surfaces that contact the food ordinarily occurs during the normal course of operations. "Food-contact surfaces" includes utensils and food-contact surfaces of equipment.
food defense The protection of food from intentional contamination by biological, chemical, physical, or radiological agents that are not reasonably likely to occur in the food supply.
food safety Assurance that food will not cause harm to the consumer when it is prepared and/or consumed according to its intended use.
Food Safety Plan A set of written documents that is based upon food safety principles; incorporates hazard analysis, preventative controls, supply-chain programs, and a recall plan; and delineates the procedures to be followed for monitoring, corrective action, and verification.
Food Safety System The system a facility implements according to the Food Safety Plan to meet its food safety needs and requirements.
Food Safety Team The group of people who are responsible for developing, implementing and maintaining the Food Safety Plan.
FSMA Food Safety Modernization Act of 2011.
Good Manufacturing Practice (GMP also cGMP) The principles, programs and practices of sanitary food production that industry must follow to provide the basic environment and operating conditions that are necessary for the production of safe, wholesome food (see 21 CFR part 117 Subpart B, formerly part 110).
HACCP Hazard Analysis Critical Control Point, a systematic approach to the identification, evaluation, and control of food safety hazards.
HACCP Plan The written document that is based upon the principles of HACCP and that delineates the procedures to be followed.
hazard Any biological, chemical (including radiological), or physical agent that has the potential to cause illness or injury.
hazard analysis The process of collecting and evaluating information on hazards associated with the food under consideration to decide which are significant and must be addressed in the HACCP plan or the Food Safety Plan.
hazard requiring a preventive control A known or reasonably foreseeable hazard for which a person knowledgeable about the safe manufacturing, processing, packing, or holding of food would, based on the outcome of a hazard analysis (which includes an assessment of the severity of the illness or injury if the hazard were to occur and the probability that the hazard will occur in the absence of preventive controls), establish one or more

preventive controls to significantly minimize or prevent the hazard in a food and components to manage those controls (such as monitoring, corrections or corrective actions, verification, and records) as appropriate to the food, the facility, and the nature of the preventive control and its role in the facility's food safety system.

known or reasonably foreseeable hazard A biological, chemical (including radiological), or physical hazard that is known to be, or has the potential to be, associated with the facility or the food.

lot The food produced during a period of time and identified by an establishment's specific code.

monitor To conduct a planned sequence of observations or measurements to assess whether control measures are operating as intended.

operating limits Criteria that may be more stringent than critical limits and are established for reasons other than food safety.

plant The facility used for or in connection with the manufacturing, processing, packaging, or holding of human food.

potential hazard A known or reasonably foreseeable biological, chemical, physical, or radiological agent that has the potential to require assignment of a preventive control, as determined by the hazard analysis' review of its severity and likelihood of occurrence.

Prerequisite Programs Procedures, including Good Manufacturing Practices (GMPs), that address the environmental and operational conditions providing the foundation for the Food Safety System.

preventive controls Risk-based, reasonably appropriate procedures, practices, and processes that a person knowledgeable about safe manufacturing, processing, packing, or holding of food would employ to significantly minimize or prevent the hazards identified under the hazard analysis that are consistent with current scientific understanding of safe food manufacturing, processing, packaging, or holding at the time of the analysis.

Preventive Controls Qualified Individual (PCQI) A qualified individual who has successfully completed training in the development and application of risk-based preventive controls at least equivalent to that received under a standardized curriculum recognized as adequate by FDA or is otherwise qualified through job experience to develop and apply a food safety system.

Process Controls Preventive controls, specifically assigned at CCPs, which provide control that is essential to prevent or eliminate a food safety hazard or reduce it to an acceptable level.

Qualified Individual (QI) A person who has the education, training, or experience necessary to manufacture, process, pack, or hold clean and safe food as appropriate to the individual's assigned duties. A QI may be, but is not required to be, an employee.

ready-to-eat food (RTE food) Any food that is normally eaten in its raw state or any other food, including a processed food, for which it is reasonably foreseeable that the food will be eaten without further processing that would significantly minimize biological hazards.

reasonably foreseeable hazard A potential biological, chemical, physical, or radiological hazard that may be associated with the facility or the food.

rework Clean, unadulterated food that has been removed from processing for reasons other than insanitary conditions or that has been successfully reconditioned by reprocessing and that is suitable for use as food.

sanitize To adequately treat cleaned food-contact surfaces by a process that is effective in destroying vegetative cells of pathogens, and in substantially reducing numbers of other undesirable microorganisms, but without adversely affecting the product or its safety for the consumer.

severity The seriousness of the effect(s) of a hazard.

significantly minimize To reduce to an acceptable level, including to eliminate.

SOP Standard operating procedure. A document that gives a step-by-step description of how a specific operation, method, or procedure is performed.

step A point, procedure, operation, or stage in the food system from primary production to final consumption.

supplier The establishment that manufactures/processes the food, raises the animal, or grows the food that is provided to a receiving facility without further manufacturing/processing by another establishment, except for further manufacturing/processing that consists solely of the addition of labeling or similar activity of a *de minimis* nature.

supply-chain-applied control A preventive control for a hazard in a raw material or other ingredient when the hazard in the raw material or other ingredient is controlled before its receipt.

validation Obtaining and evaluating scientific and technical evidence that a control measure, combination of control measures, or the food safety plan as a whole, when properly implemented, is capable of effectively controlling the identified hazards.

verification The application of methods, procedures, tests, and other evaluations, in addition to monitoring, to determine whether a control measure or combination of control measures is or has been operating as intended and to establish the validity of the food safety plan.

water activity (a_w) A measure of the free moisture in a food, namely the quotient of the water vapor pressure of the substance divided by the vapor pressure of pure water at the same temperature.

(See also 21 CFR Subpart A Part 117.3 for additional definitions.)

www.ingramcontent.com/pod-product-compliance
Lightning Source LLC
Chambersburg PA
CBHW072234250125
20788CB00020B/292
9781119258070